CAD of Microstrip Antennas for Wireless Applications

DISCLAIMER OF WARRANTY

For a complete listing of the *Artech House Antenna Library*,
turn to the back of this book.

CAD of Microstrip Antennas for Wireless Applications

Robert A. Sainati

Artech House
Boston • London

Library of Congress Cataloging-in-Publication Data

Sainati, Robert A.

CAD of microstrip antennas for wireless applications/Robert A. Sainati

p. cm.

Includes bibliographical references and index.

ISBN 0-89006-562-4 (alk. paper)

1. Microstrip antennas–Design–Data processing. 2. Computer-aided design. 3. Wireless communication systems. I. Title.

TK7876.S24 1996

621.381'331–dc20 96-13364

CIP

British Library Cataloguing in Publication Data

Sainati, Robert A.

CAD of microstrip antennas for wireless applications

1. Computer-aided design 2. Microstrip antennas - Design and construction - Data processing

I. Title

621.3'824'0285

ISBN 0-89006-562-4

685 Canton Street

Norwood, MA 02062

International Standard Book Number: 0-89006-562-4

Library of Congress Catalog Card Number: 96-13364

10 9 8 7 6 5 4 3 2 1

Contents

Preface

Until the early 1990s, RF and microwave technologies were primarily used for military applications. Often these applications demanded state-of-the-art performance along with the ability to operate in virtually any imaginable environment. Since the beginning of the current decade, there has been a rapid shift away from military toward commercial systems. An explosion of new commercial products utilizing *wireless* technology has come onto the scene. Typically, commercial products are developed under significantly different circumstances than military systems. Cost is usually paramount; the product must be easy to use; product size and aesthetics are often important; and in many cases, performance does not need to be optimum. The fast pace of most commercial markets demands a change from the lengthy design cycle required to develop high-performance military systems to very short product development times.

Because of the short design cycle times, engineers in commercial organizations are often called upon to work in technical areas where they may not have a strong background. Most products incorporating RF or microwave technology require some sort of antenna, yet traditionally antennas are only given cursory coverage in most electrical engineering curriculums. First-generation commercial products have been able to use simple antenna structures such as monopoles and basic reflectors. This is, in part, because they are easy to design and therefore low in cost to develop. However, as customers demand ''smarter'' and smaller products, these simple antennas may not be satisfactory anymore. A competitive advantage can often be gained by incorporating an antenna that can be made integral to the product housing or that replaces the reflector with a flat plate. In addition to smaller size and better aesthetics, this antenna may be required to provide improved performance.

Microstrip antennas have several features that make them ideal for commercial applications. Based upon printed circuit technology, they are very inexpensive to produce.

Production of large quantities is easy. Their inherent low profile allows them to be mounted directly on product housings. Often the antenna can be integrated on the board containing the rf or microwave circuits. The purpose of this book is to help develop an understanding of the operation of microstrip antennas and supply the tools necessary for design work. It is aimed at the nonspecialist. Some theory is presented but mainly to give the reader physical insight that I believe is necessary for successful design. Most of the design tools are given in the form of computer programs. Each program is intended to cover either the design of a particular antenna or some aspect of the design. Hopefully after reading this book, an engineer will have the necessary background to develop microstrip antennas for various applications.

A disk that contains the computer programs is supplied along with the book. The programs are all stand-alone, executable codes that typically analyze and solve a particular problem. There is no menu or Windows-type interface that is wrapped around the programs. You have to pick the appropriate program and run it to perform the analysis or design. This may be disappointing to some, but I did it in this manner for a couple of reasons. First, for better or worse, this approach follows my personal design style. Over the years, I have written programs as part of a design effort; that is, any given program concentrated on solving a particular aspect of the development. In some cases, I modified the programs later on to increase their applicability or to make their use easier. This book and accompanying disk represent a compilation of those efforts over the years. The second reason is that I am of the generation that saw the introduction of the computer into electrical engineering. I certainly recognize the importance and value of using a computer in design work, but to me it has always been a tool, not an end in itself. I appreciate nice interfaces as much as anyone, but I have never considered them an absolute necessity.

We are all the product of the people with whom we have interacted in our lives. I would like to acknowledge some of the many people who have been a part of my professional career. First there are those from whom I learned, namely Professors P. D. Coleman and P. E. Mast at the University of Illinois at Urbana and Professor C. W. Schultz at the University of Connecticut in Storrs. There are those with whom I have had the privilege of working, particularly T. Bird, G. James, and M. Wilkes at the CSIRO Division of Radiophysics, Sydney, Australia; S. Fraasch, M. Komen, and B. Carlyon at Alliant Techsystems, Hopkins, MN; J. Pozgay at Raytheon Missile Systems Division, Bedford, MA; and J. Comiskey, D. Fessenden, C. Foster, and T. Vuono at the Naval Underwater Systems Center, New London, CT. Thanks to S. Barker of CSIRO Division of Radiophysics for helping me create some Postscript files. Special thanks goes to B. Herrington at Williams Sound, Eden Prairie, MN, for generating the Smith chart plotting routine used in many of the programs. Finally, and most importantly, I dedicate this book to Jill and Lynn—they have given my life meaning.

Chapter 1
Introduction

RF and microwave technologies are rapidly finding their way into commercial and consumer products. These technologies have traditionally been used in specialist-designed military systems. To be competitive in the commercial marketplace, design cycle times must be short and low-cost. Engineers often find themselves in the position of having to perform design work in unfamiliar areas such as antennas and RF circuits. The aim of this book is to help engineers develop an understanding of microstrip antennas and their design using supplied computer programs.

As commercial systems evolve, more and more demands on performance will occur. Still the product must be easy to use, low-cost, and physically small. The simple monopole and reflector antennas of today's products may no longer be sufficient. Microstrip antennas provide the technology necessary to meet the demands of future products. This chapter introduces the microstrip antenna concept along with advantages, disadvantages, and applications. Since antennas are electromagnetic devices, there is also a brief review of Maxwell's equations and antenna terms.

1.1 PURPOSE AND SCOPE OF THE BOOK

During the past few years there has been an explosion in commercial applications involving RF and microwave systems. Industrial applications such as satellite data transfer, vehicle tracking, and personal paging have been among the first to be developed. Another early application is the mobile telephone. The future will see even further penetration of RF and microwave systems into both the workplace and personal lives. *Direct broadcast television* (DBS) and *digital audio broadcast* (DAB) allow reception of entertainment virtually anywhere. Wireless *local area networks* (LANs) and *personal communications systems* (PCS) provide untethered data transfer and communications. The intelligent vehicle highway of the future will guide us through traffic jams and tell us about services along the way. Finally, systems using GPS and other techniques not only tell us where we are going but where we should be going.

Most of these commercial systems must be low-cost, easy to use, small in size, and rugged to achieve wide acceptance. Low cost demands easily produced components. The drive for smaller systems pushes not only integrated circuit technology but also antenna technology. Small, conformal antennas are aesthetically pleasing and increase product ruggedness by avoiding antenna breakage. Many applications currently use straightforward antennas (for example, monopoles and reflectors) partly because they are well understood and relatively easy to design. As customers demand smarter, smaller, lower cost products, more innovative antennas will be required. Microstrip antennas are ideal for these applications. They are very thin and, if made on flexible substrates, can be conformal. They are compatible with IC technology in the sense that they readily interface with IC interconnects. They can even be made part of the chip. Most importantly, microstrip antennas are manufactured with printed circuit techniques and, therefore, are very low in cost.

Antenna design has traditionally been the realm of those specially trained in that discipline. One facet of the emerging commercial applications is the need for short design cycle times. This is in response to rapidly changing marketplaces. The design engineer is asked to participate in all aspects of the product design. Often they have only had a cursory exposure to antennas. There may not be time to "come up to speed," and consultants may be hard to find or too costly. This in part is why monopoles and simple reflectors are used in today's systems. The purpose of this book is to provide the nonspecialist with an understanding of microstrip antennas and some tools, mostly in the form of computer programs, that are necessary for their design.

The book presents a description of microstrip antennas and their design, with the emphasis on developing insights into microstrip antenna operation. Developing an understanding of antenna behavior is a key element for successful design. Many of the mathematical models used to analyze the antennas are based on various physical models. A discussion of these physical models helps give an understanding of antenna operation. The book then describes the design of several types of single-element radiators and arrays of elements. In most cases computer programs are included that implement the design procedure. Examples are presented to show how the programs are used. This book is not intended to provide an exhaustive treatment of antennas or even a complete discussion of microstrip antennas. Its primary goal is to help engineers design the most commonly used microstrip antennas.

1.2 BRIEF HISTORY AND DESCRIPTION OF MICROSTRIP RADIATORS

The idea for microstrip radiators dates back to the early 1950s and a concept proposed by Deschamps [1]. Several years later Gutton and Baissinot [2] patented a microstrip-based antenna. In spite of the publication of the concept, not much activity in microstrip antenna development occurred over the next 15 years or so except for some unpublished work by Kaloi at the U.S. Navy Missile Test Range in California. This was in part due to the lack of good microwave substrates. Also at that time more interest was focused on stripline circuits and antennas as thinner, lower cost alternatives to waveguide components. The need for thin, conformal antennas for missiles and spacecraft started the rapid development of microstrip antennas in the early 1970s. Some of this early work is presented in [3–6].

Microstrip antennas are an extension of the microstrip transmission line. First proposed in 1952 by Grieg and Englemann [7], microstrip was overshadowed by stripline as an emerging printed circuit media for microwave components. Stripline supports a *transverse electromagnetic* (TEM) wave, making the analysis and design of components easier. Also the lack of active semiconductor components meant that most circuits were passive structures. Stripline's sandwich structure is well suited for these circuits but not so for active devices that require space for bonding wires. Microstrip became popular with the advent of microwave transistors and other active devices. Complete circuits such as amplifiers, phase shifters, and mixers could be realized in hybrid form thanks to microstrip.

The basic microstrip structure is given in Figure 1.1. The bottom surface of a thin dielectric substrate is completely covered with metalization that serves as a ground plane. The metalization is usually copper that has been electrodeposited or rolled on. With the former, the copper is chemically deposited on the substrate. For the latter, a thin copper sheet is attached by an adhesive. Typical thickness for the dielectric, h, range from 0.005 cm (0.002 in) to 0.635 cm (0.250 in) or greater. For microwave circuits substrate materials such as alumina, quartz, and glass-loaded PTFE are used. The copper thickness, t, is usually about 0.001778 cm (0.0007 in) to 0.003556 cm (0.0014 in). The microstrip line is formed by etching away all of the metal on the top surface except for a strip of width w. The substrate is electrically characterized by its relative dielectric constant, ϵ_r, and a loss tangent, tan δ.

The strip and ground plane form a transmission line that guides energy with a quasi-TEM wave. The substrate thickness is a very small fraction of a wavelength about 0.02λ

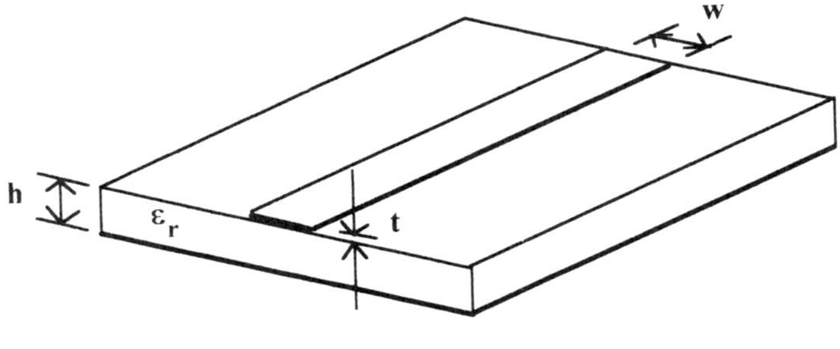

Figure 1.1 Microstrip transmission line configuration.

or less. As long as the physical dimensions and relative dielectric constant remain constant, virtually no radiation occurs. The microstrip will radiate by suitably shaping the microstrip line as in Figure 1.2, which shows a rectangular microstrip patch antenna. The antenna is shown being fed with a microstrip line, though other feeding techniques are possible. The discontinuity introduced by the rapid change in line width at the junction between the feed line and patch radiates. Radiation also occurs at the other end of the patch where the metalization abruptly ends. In actuality, the sides of the patch radiate, but the level is usually well below that of the other edges.

All microstrip antennas are essentially discontinuities, usually in some physical dimension, that radiate. The shape of the discontinuity is chosen so that well-defined radiation patterns result. A properly chosen shape also results in an impedance versus frequency behavior that approximates a parallel RLC circuit. The operating frequency is determined when the impedance becomes purely real. Many shapes, including rectangles, circles, and triangles, have been used. There is another way to view patch radiation as coming from a resonator that leaks away power. The above shapes can be treated as resonators with operating frequencies determined by the resonant modes. These physical models for patch radiation will be discussed in Chapter 2.

1.3 ADVANTAGES AND DISADVANTAGES OF MICROSTRIP ANTENNAS

The interest in microstrip antennas arises from several desirable characteristics. The first is low profile. With microstrip, antennas become a fraction of an inch thick, which makes them easy to incorporate into a vehicle shell or a package lid. If the substrate is flexible, conformal antennas are possible. Instead of protruding out somewhere or requiring a separate cavity, the antenna now becomes an integral part of the housing or package. Light weight is another advantage. Microstrip antennas are extremely low in cost to manufacture. Machining of complicated metal parts is not required. Etching is done with

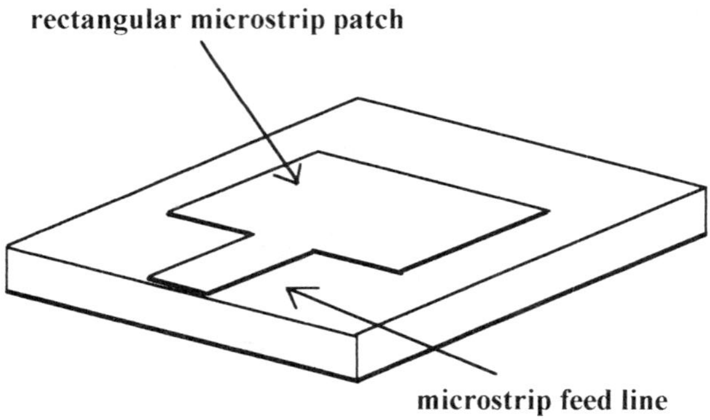

Figure 1.2 Rectangular microstrip patch antenna configuration.

the standard photolithographic processes. The accuracy of etching processes ensures uniformity of parts over a production run. Production of large quantities becomes quite easy. Microstrip antennas are highly integrated. The antenna circuit board may include the radiating element(s) and all the feed network. Antennas can be placed on the same substrate as other circuitry, further increasing the level of integration. They have even been incorporated on integrated circuit chips.

Like any technology, there are disadvantages. The major limitation is the narrow bandwidth. The impedance versus frequency behavior of virtually all microstrip patches limits the operating frequency range. Typical patch quality factors, Q, are around 50 to 75. Bandwidth is a function of Q and the acceptable mismatch. Defining bandwidth as the frequency range over which the standing wave ratio is 2.00:1 or less, patch bandwidths are 1% to 5%. Of course, the bandwidth is a function of the tolerable mismatch, so it can vary depending upon the application. The antenna quality factor is inversely proportional to antenna size in terms of the wavelength. Small antennas, like patches, inherently have high quality factors and narrower bandwidths. Large arrays of microstrip antennas have several disadvantages. The narrow patch bandwidth makes the design of very low sidelobe antennas very difficult. Also spurious radiation from microstrip feed lines can corrupt patterns. Finally, microstrip feed line losses result in relatively low efficiencies.

1.4 APPLICATIONS OF MICROSTRIP ANTENNAS

The first applications for microstrip antennas were those requiring thin, conformal antennas. The telemetry and communications antennas on missiles are often microstrip. Small arrays of microstrip radiators are used for radar altimeter antennas. Other aircraft-related applications include antennas for telephone and satellite communications. Microstrip arrays have been used on satellite imaging systems such as SEASAT and SIR-A. Because of the size and efficiency requirements for these antennas, special feeding techniques were used to minimize losses. Patch antennas have been used on communications links between ships or buoys and satellites such as GOES and SMS. Smart weapons systems use microstrip antennas because of their thin profile and low cost. Today GPS systems are proving to be a major user of microstrip antennas. For details on some of these applications see [8].

1.5 REVIEW OF ELECTROMAGNETIC THEORY

Although this book will not cover the electromagnetic analysis of microstrip antennas in great detail, it is important to have an understanding of some of the basic concepts. The concepts of electric and magnetic fields arose as an outcome of experiments studying the forces on static (dc) charges and currents. For situations that are static, the electric and magnetic fields are uncoupled and exist independent of each other. Electric fields describe the interaction between charges. Phenomena involving currents are characterized with magnetic fields. Later experiments showed that with time-varying (ac) quantities, electric and magnetic fields become coupled and exist simultaneously. In 1873, Maxwell consolidated the previous work of Coulomb, Ampere, Faraday, and others into a set of equations that essentially predicts all electromagnetic phenomena at the macroscopic level. Maxwell also introduced the concept of displacement current, which eventually led to the discovery of electromagnetic waves. Excellent introductory discussions of electromagnetic theory are contained in [9,10].

Fields are vector quantities; that is, they are characterized by a magnitude and direction. The magnitude of static fields can vary with position. Time-varying fields are functions of position and time. Using Cartesian coordinates, an arbitrary electric field can be expressed as

$$\overline{E} = E_x(x, y, z, t)\overline{\mathbf{x}} + E_y(x, y, z, t)\overline{\mathbf{y}} + E_z(x, y, z, t)\overline{\mathbf{z}} \tag{1.1}$$

where the overbar denotes a vector quantity. E_x, E_y, and E_z are the x-, y-, and z-components of the vector. In general, as indicated in (1.1), they are functions of all three spatial variables, x, y, and z, and time, t. $\overline{\mathbf{x}}$, $\overline{\mathbf{y}}$, and $\overline{\mathbf{z}}$ are unit vectors along the x-, y-, and z-axes, respectively. Most applications involve sinusoidal time dependence where a steady state has been reached. In this situation, the use of phasors becomes very convenient. Assuming a sinusoidal field with only an x-component, (1.1) becomes

$$\overline{E} = E_x(x, y, z)\cos(\omega t + \phi)\overline{\mathbf{x}} \tag{1.2}$$

where E_x is now only a function of x, y, and z, ω is the radian frequency, and ϕ is the phase at $t = 0$. Equation (1.2) can be expressed in phasor form as

$$\overline{E} = E_x(x, y, z)e^{j\phi}\overline{\mathbf{x}} \tag{1.3}$$

which suppresses the (known) time dependence. In most instances, this is the manner in which fields will be expressed in this book. To convert from a phasor back to an instantaneous field, multiply the phasor by $e^{j\omega t}$ and then take the real part

$$\overline{E} = \mathrm{Re}\{[E_x(x, y, z)e^{j\phi}]e^{j\omega t}\}\overline{\mathbf{x}} \tag{1.4}$$

The factor $e^{j\omega t}$ is understood whenever phasors are used and will be dropped.

Electromagnetic fields obey a set of equations named after Maxwell. There are two forms of Maxwell's equations. When considering fields in a localized region in space, the point or differential form is appropriate. For this case, Maxwell's equations in phasor form are written as

$$\nabla x\overline{E} = -j\omega\overline{B} \tag{1.5}$$

$$\nabla x \overline{H} = j\omega \overline{D} + \overline{J} \tag{1.6}$$

$$\nabla \cdot \overline{D} = \rho \tag{1.7}$$

$$\nabla \cdot \overline{B} = 0 \tag{1.8}$$

where $\overline{E}$ is the electric field intensity, in volts per meter (V/m); $\overline{B}$ is the magnetic flux density, in webers per square meter (Wb/m^2); $\overline{H}$ is the magnetic field intensity, in amperes per meter (A/m); $\overline{D}$ is the electric flux density, in coulombs per square meter (C/m^2); $\overline{J}$ is the electric current density, in amperes per square meter (A/m^2); and ρ is the electric charge density, in coulombs per cubic meter (C/m^3). Also $\nabla \times$ and $\nabla \cdot$ denote the curl and divergence of a vector, respectively. In free space, $\overline{D} = \epsilon_0 \overline{E}$ and $\overline{B} = \mu_0 \overline{H}$, where $\epsilon_0 = 8.854 \times 10^{-12}$ F/m is the permittivity (of free space) and $\mu_0 = 4\pi \times 10^{-7}$ H/m is the permeability. Usually the current, $\overline{J}$, and charge density, ρ, are considered to be the sources of the fields. When talking about fields radiated from an antenna, the $\overline{E}$- and $\overline{H}$-fields are of most interest.

Since current is the flow of charge, $\overline{J}$ is related to ρ. This relationship is expressed by the equation of continuity

$$\nabla \cdot \overline{J} = \frac{\partial \rho}{\partial t} \tag{1.9}$$

which is a restatement of conservation of charge. At a particular point, any change in charge with time, as indicated by the derivative, must be caused by current flowing into or out of the point as found from the divergence.

With problems concerned with field behavior in a region of space, the integral forms of Maxwell's equations are more applicable. In phasor form, they are

$$\int_c \overline{E} \cdot d\overline{l} = -j\omega \iint_S \overline{B} \cdot d\overline{S} \tag{1.10}$$

$$\int_c \overline{H} \cdot d\overline{l} = \iint_S \overline{J} \cdot d\overline{S} + j\omega \iint_S \overline{D} \cdot d\overline{S} \tag{1.11}$$

$$\iint_S \overline{D} \cdot d\overline{S} = \iiint_V \rho \cdot dV \tag{1.12}$$

$$\iint_S \overline{B} \cdot d\overline{S} = 0 \tag{1.13}$$

Equation (1.10) is Faraday's Law. For a loop antenna, the left-hand side integral gives the voltage induced at the terminals due to the right-hand side integral of the time derivative (as given by the factor $j\omega$) of the magnetic flux density passing through the loop. Equation (1.11) is a modified version of Ampere's Law. For static fields ($\omega = 0$), the integral of the magnetic field around a closed path yields the current flowing through the area enclosed by the path. With a capacitor, no charge flows between the plates ($\overline{J} = 0$), so (1.11) shows that the capacitor "current" is due to the integral of the time derivative of a time-varying electric field across the plates. Since electric flux lines begin and end on charges, (1.12) indicates that the total electric flux density passing through a closed surface is equal to the total charge contained inside the enclosed volume. There are no known magnetic charges; therefore (1.13) says that magnetic flux density lines close upon themselves.

When electric and magnetic fields exist inside material bodies, they interact with the electrons, atoms, and molecules of the material. An applied electric field polarizes atoms and molecules by displacing the negatively charged electrons from the positive nucleus, creating an additional component to the electric flux density. For most media, the electric flux density is

$$\overline{D} = \epsilon_0 \overline{E} + \overline{P} \tag{1.14}$$

where $\overline{E}$ is the applied electric field and $\overline{P}$ is the electric polarization vector. When the medium is linear, $\overline{P}$ is related to $\overline{E}$ by

$$\overline{P} = \epsilon_0 \chi_e \overline{E} \tag{1.15}$$

where χ_e is the electric susceptibility and is usually complex. Substituting (1.15) into (1.14), we can define the complex permittivity of the medium as

$$\epsilon = \epsilon' - j\epsilon'' = \epsilon_0(1 + \chi_e) \tag{1.16}$$

Normally, the real part of (1.16) is divided by ϵ_0 and called the relative dielectric constant

$$\epsilon_r = \epsilon'/\epsilon_0 \tag{1.17}$$

This is commonly specified on material data sheets. The imaginary part of ϵ accounts for losses in the material. The ratio of imaginary to real parts is called the loss tangent and is denoted

$$\tan\delta = \epsilon''/\epsilon' \tag{1.18}$$

This is also given as a material specification. Usually $\tan\delta$ should be as small as possible, such as less than 0.001 or so.

There are magnetic materials where the applied magnetic field induces magnetic dipole moments that contribute to the magnetic flux density. Following a similar argument as used for the permittivity, $\overline{B}$ is related to $\overline{H}$ according to the equation

$$\overline{B} = \mu_0(1 + \chi_m)\overline{H} = \mu\overline{H} \tag{1.19}$$

where μ is the material permeability and χ_m is the (complex) magnetic susceptibility. Like the permittivity, μ can be complex, $\mu = \mu' - j\mu''$, if there are losses. The relative permeability is simply

$$\mu_r = \mu'/\mu_0 \tag{1.20}$$

For most media encountered in this book, $\mu = \mu_0$; that is, they are nonmagnetic materials. Therefore $\mu_r = 1$.

Finally in some materials, like metals, there are electrons that can freely react to an applied electric field. When an electric field is impressed in such a material, the charges move in the direction of the applied field, resulting in current flow. The relationship between resulting current density and applied field is expressed as

$$\overline{J} = \sigma\overline{E} \tag{1.21}$$

where σ is referred to as the conductivity. For many metals, which are good conductors, the conductivity is on the order of 1 to 5×10^8 S/m.

1.6 REVIEW OF ANTENNA PARAMETERS

The characteristics of interest for an antenna are its radiation patterns, beamwidth, directivity/gain, polarization, impedance, efficiency, and bandwidth. References [11] and [12] present definitions for these quantities. The radiation of a small dipole will be used to illustrate some of the parameters.

The electromagnetic fields associated with a small dipole consists of three components that in spherical coordinates are E_r, E_θ, and H_ϕ. For a small dipole of length dl lying along the z-axis at the origin of the coordinate system, the fields are given by [13]

$$E_r = \frac{j\eta_0 Idl}{2\pi k_0}\cos(\theta)\left[\frac{jk_0}{r^2} + \frac{1}{r^3}\right]e^{-jk_0r} \tag{1.22}$$

$$E_\theta = -\frac{j\eta_0 Idl}{4\pi k_0}\sin(\theta)\left[-\frac{k_0^2}{r} + \frac{jk_0}{r^2} + \frac{1}{r^3}\right]e^{-jk_0r} \tag{1.23}$$

$$H_\phi = \frac{Idl}{4\pi}\sin(\theta)\left[\frac{jk_0}{r} + \frac{1}{r^2}\right]e^{-jk_0r} \tag{1.24}$$

where η_0 is the wave impedance of free space, denoted $\sqrt{\mu_0/\epsilon_0}$; I is the current on the dipole; k_0 is the wavenumber, denoted $2\pi/\lambda_0$; λ_0 is the free-space wavelength; and r is the radial distance from the dipole. θ is measured from the dipole axis (which is also the z-axis). Because of symmetry, the dipole fields do not depend on the other spherical coordinate angle ϕ, which is measured from the x-axis. The electric and magnetic fields contain terms that vary as $1/r^2$ and $1/r^3$. It can be shown that these terms are associated with energy storage around the antenna. The terms having a $1/r$ dependence become dominant at large distances and represent power radiated by the antenna.

The terms that vary as $1/r^2$ and $1/r^3$ are called the reactive field or near-field components. The near-field components are dominant in a region around the antenna defined by a sphere of radius $r \leq 0.62 \cdot \sqrt{D^3/\lambda_0}$, where D is the largest dimension of the antenna. The region is called the reactive near-field zone [14]. This definition is not based upon a rigorous analysis but is rather an adopted convention that seems to apply to most antennas. The terms that vary as $1/r$ are called the far-field components. The spherical region defined by $r \geq 2D^2/\lambda_0$ is where the far-field components dominate and is the far-field or Fraunhofer zone. The region in between is the radiating near-field or Fresnel zone.

The far-field components of the small dipole are

$$E_\theta = \frac{j\eta_0 k_0 Idl}{4\pi}\sin(\theta)\frac{e^{-jk_0 r}}{r} \tag{1.25}$$

$$H_\phi = \frac{jk_0 Idl}{4\pi}\sin(\theta)\frac{e^{-jk_0 r}}{r} \tag{1.26}$$

If converted from phasor form back to instantaneous fields, E_θ and H_ϕ would be in-phase. They are also spatially orthogonal, and their ratio is the free-space wave impedance. These characteristics are common for far fields from any antenna. The dipole electric field always lies in a straight line that *points* along the θ-direction. The dipole is said to be linearly polarized. Polarization will be discussed later in this section.

When r is large enough to be in the far field, the radiated fields share the same r-dependency. Their variation with the polar angles θ and ϕ is therefore independent of r. This is not true where the reactive components are significant. The radiation pattern is thus defined as the field variation with angle(s) in the far field. For the dipole, the pattern is given by the function $\sin(\theta)$. With a linearly polarized antenna, the pattern in a plane parallel to the electric field is called the E-plane pattern. Since the dipole fields are independent of ϕ, any plane containing the dipole defines the E-plane pattern. Conversely, the H-plane pattern lies in a plane perpendicular to the electric field. The (x-y)-plane defines the H-plane pattern of the dipole.

The radiation pattern is usually plotted in decibels, that is $20 \cdot \log[E(\theta, \phi)]$, versus angle. Figure 1.3 shows E- and H-plane patterns for the dipole. The E-plane pattern is the familiar "figure 8" pattern of a dipole. The H-plane pattern is considered to be omnidirectional since it does not vary with angle.

(a)

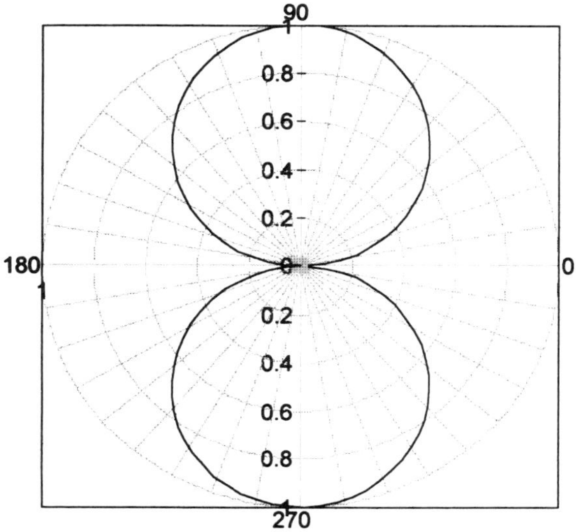

(b)

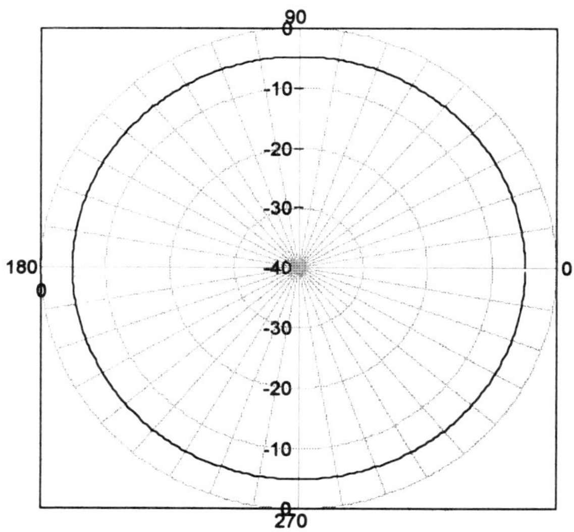

Figure 1.3 Small dipole antenna patterns: (a) E-plane and (b) H-plane.

The radiation patterns of a small dipole are relatively simple. With larger antennas, like parabolic reflectors or arrays of dipoles, the patterns display more features that often must meet design specifications. Figure 1.4 presents the pattern for a ten-element array. Most of the radiation is concentrated within the peak part of the pattern denoted by the main beam. An important parameter of the main beam is its width, which is called the 3-dB beamwidth. It is the angular width of the main beam between points that are 3 dB (half-power) down from the pattern peak. Obtaining a particular 3-dB beamwidth is usually a major design goal. There are also secondary peaks called sidelobes that are specified in terms of decibels down from the main beam. One sidelobe (usually that next to the main beam) has a higher magnitude than the others. This is referred to as the peak sidelobe. Often there is a maximum value for this peak sidelobe that must be met for a particular application. The beamwidth and peak sidelobe levels are interrelated parameters, as will be discussed in Chapter 6. To decrease the peak sidelobe, the beamwidth must widen.

In most situations, antennas are reciprocal devices. Any property ascribed to an antenna when used to transmit also applies if used to receive. As a consequence, the radiation pattern is the same whether the antenna is transmitting or receiving. The same is true for other properties such as directivity, polarization, and impedance.

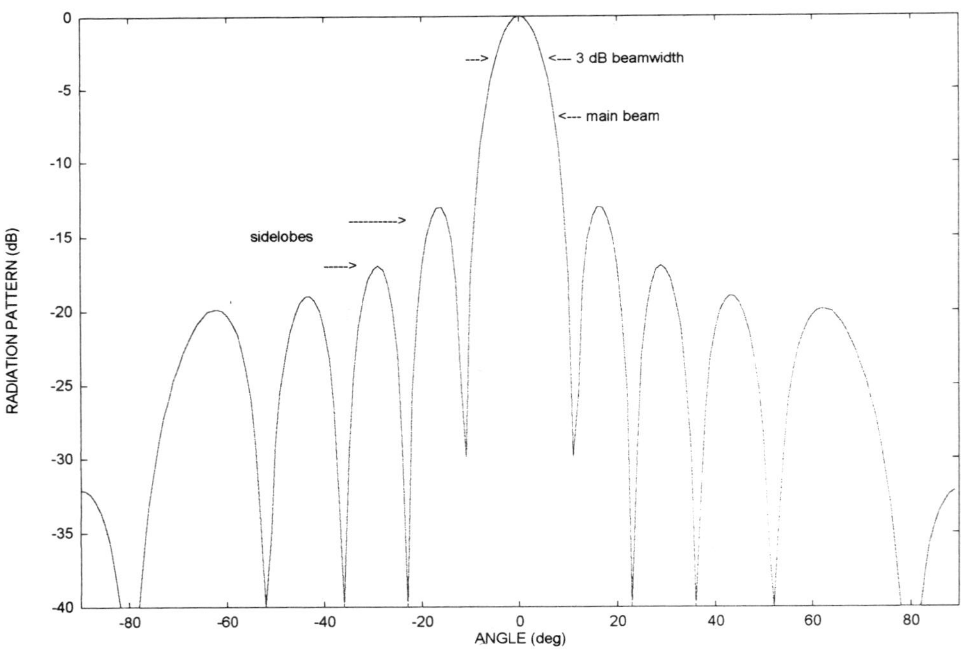

Figure 1.4 Radiation pattern for a ten-element array.

The electromagnetic waves radiated by an antenna carry power away from the antenna. The power density of a wave is given by Poynting's vector. For sinusoidal time-varying fields, Poynting's vector is

$$\overline{P_{avg}} = \frac{1}{2}\text{Re}\{\overline{E} \times \overline{H^*}\} \tag{1.27}$$

The factor 1/2 indicates that peak values of the fields are used. The asterisk denotes the complex conjugate of the magnetic field. $\overline{P_{avg}}$ is the time-averaged power density in watts per square meter (W/m^2). It is a vector whose direction points in the direction of wave propagation. In the far field, (1.27) is predominately real, indicating power flow away from the antenna. The total (average) power radiated is given by the integral of (1.27) over an appropriate surface, S as

$$P_{rad} = \frac{1}{2}\int\!\!\int_S \text{Re}\{\overline{E} \times \overline{H^*}\} \cdot d\overline{S} \tag{1.28}$$

With an antenna in free space, the integration surface is a complete sphere (where usually, $d\overline{S} = r^2\sin(\theta)d\theta d\phi$ and the range of integration goes from $\theta = 0$ to π and $\phi = 0$ to 2π). For antennas on or over an infinite ground plane, the integral is performed over a hemisphere.

Let P_{avg} be the magnitude of the Poynting vector. Since the far fields have a $1/r$ dependency, the quantity

$$U = r^2 P_{avg} \tag{1.29}$$

represents the power radiated per unit solid angle [14]. U is called the radiation intensity and, in general, is a function of the angles θ and ϕ. In the far field of any antenna, $|\overline{H}| = |\overline{E}|/\eta_0$, (see (1.25) and (1.26)) and U becomes

$$U = \frac{r^2}{2\eta_0}[|E_\theta(\theta, \phi)|^2 + |E_\phi(\theta, \phi)|^2] \tag{1.30}$$

where, in the general case, the antenna radiates both E_θ and E_ϕ components.

The pattern in Figure 1.4 shows that the ten-element array concentrates most of its radiated power in the main beam. The ability of an antenna to focus radiation is an important parameter and is characterized by the term directivity. By definition [14] the directivity is the ratio of the maximum radiation intensity to the radiation intensity of a reference antenna. With the array, the maximum intensity would occur at the peak of the main beam at 0 degrees (see Figure 1.4). The reference antenna is chosen to be an isotropic radiator, which is one that radiates equal power in all directions (one does not exist in reality). From the definition

$$D = \frac{U_{max}}{U_0} \tag{1.31}$$

where D is the directivity, U_{max} is the maximum radiation intensity, and U_0 is the intensity of an isotropic radiator.

For an isotropic radiator, the power is the same in any direction; therefore, and by definition, the Poynting vector U is constant at all angles. Let the isotropic radiation intensity be U_0. The total radiated power (1.28) becomes

$$P_{rad} = 4\pi U_0 \tag{1.32}$$

Using this, the directivity can be expressed as

$$D = \frac{4\pi U_{max}}{P_{rad}} \tag{1.33}$$

where P_{rad} is the total radiated power. By assuming the same total power for both antennas, P_{rad} can be calculated using the fields of the actual antenna, and one does not have to worry about a fictitious isotropic source. The directivity is a measure of the directional properties of an antenna compared to those of an isotropic antenna. Since an isotropic antenna has "no" directivity, D will always be greater than one. There is a relationship between beamwidth and directivity; that is, the directivity increases as the beamwidth narrows. This relationship is covered in Chapter 6.

Gain is a quantity that is related to directivity. It is defined as the ratio of the maximum radiation intensity to the power input, P_{in}, to the antenna and is denoted

$$G = \frac{4\pi U_{max}}{P_{in}} \tag{1.34}$$

The only difference between gain and directivity lies in the power used in the denominator. The power radiated equals the input power minus any power dissipated in the antenna. For an antenna with no losses, $P_{rad} = P_{in}$ and gain equals directivity. Virtually all antennas have some loss mechanisms usually in the form of conductor and dielectric losses. As a result, less power is radiated than input. The ratio of radiated to input power is expressed as

$$P_{rad} = eP_{in} \tag{1.35}$$

where e is the efficiency, which lies in the range $0 \leq e \leq 1$. Usually the efficiency is expressed as a percentage. From (1.33) to (1.25)

$$G = eD \tag{1.36}$$

When losses are present, the gain is lower than the directivity. Gain is typically expressed in decibels as $10 \cdot \log(G)$. Because of reciprocity, the directivity and gain have the same values for a receiving antenna. In this case, P_{rad} becomes the total received power and P_{in} is the power delivered to a receiver.

The gain is a useful quantity for systems studies. Given a transmitter power of P_{t} watts, an antenna in free space with gain G_{t} produces an electromagnetic field whose Poynting vector magnitude is

$$P_{\text{avg}} = \frac{G_{\text{t}} P_{\text{t}}}{4\pi r^2} \tag{1.37}$$

at a distance r away. As shown in [15], the power delivered by a receiving antenna with gain G_{r} when immersed in a field with a power density P_{avg} can be expressed as

$$P_{\text{rec}} = \frac{\lambda_0^2}{4\pi} G_{\text{r}} P_{\text{avg}} \tag{1.38}$$

So the gain helps determine both the power radiated and received. Equations (1.37) and (1.38) may be combined to give

$$P_{\text{rec}} = \frac{\lambda_0^2 P_{\text{t}}}{(4\pi r)^2} G_{\text{t}} G_{\text{r}} \tag{1.39}$$

which is a version of the Friis transmission formula. Chapter 5 of [15] presents a complete discussion of the Friis formula as its application.

In general an antenna will radiate an electric field having both θ- and ϕ-components. Let the electric field be

$$\overline{E} = E_\theta(\theta, \phi)\overline{\theta} + E_\phi(\theta, \phi)e^{j\alpha}\overline{\phi} \tag{1.40}$$

Converting to instantaneous fields,

$$\overline{E} = E_\theta(\theta, \phi)\cos(\omega t)\overline{\theta} + E_\phi(\theta, \phi)\cos(\omega t + \alpha)\overline{\phi} \tag{1.41}$$

At a fixed location in space, the figure traced out with time by the electric field, as given in (1.41), defines the antenna polarization. Consider the case where $\alpha = 0$. The fields are in time phase, and their vector sum always lies along a line that makes an angle

$$\delta = \tan^{-1}\left[\frac{E_\theta(\theta, \phi)}{E_\phi(\theta, \phi)}\right] \tag{1.42}$$

with respect to the ϕ-axis. Figure 1.5(a) shows a case where $E_\theta(\theta, \phi) < E_\phi(\theta, \phi)$. The fields are propagating outwardly in a direction normal to the plane of the paper. At $t = 0$ the two components are positive and at their maximum values as shown by the bold vectors. They combine to form a total field vector as shown in the figure. As t increases, the cosine becomes smaller and the magnitude of the total vector shrinks. When $\omega t = 90$ degrees, both components are identically zero. Beyond this time the fields *point* along the negative axes. The total field enters the third quadrant, increasing in magnitude

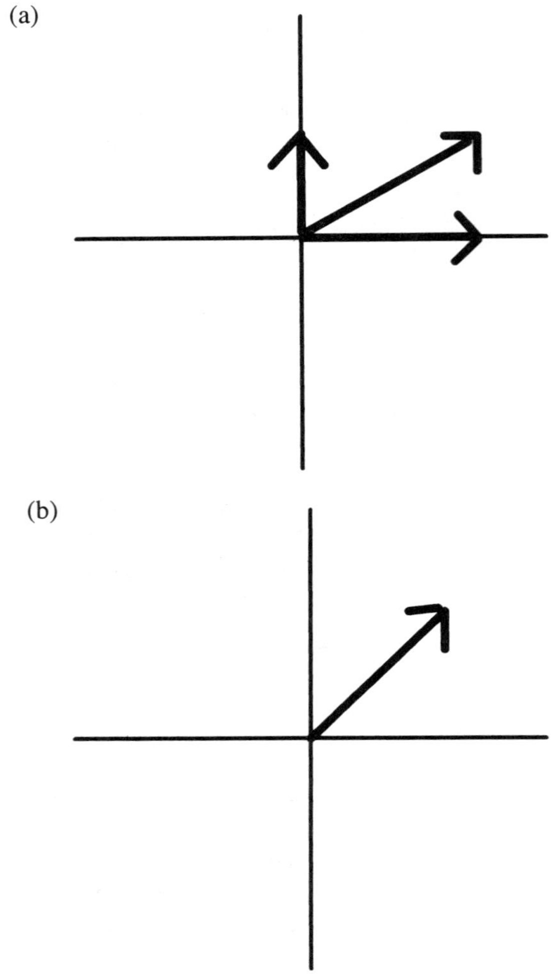

Figure 1.5 Far-field antenna with (a) linear and (b) circular polarization.

until the maximum is reached at ωt = 180 degrees, then decreasing with time, and so on. The total field vector traces out a line; hence, the antenna is considered to be linearly polarized. In most cases, in order to simplify the analysis of linearly polarized antennas the coordinate system is set up so that the radiated electric field lies along one of the axes.

The other case of interest occurs when $\alpha = \pm 90$ degrees and the two components are of equal magnitude. Then (1.41) becomes

$$\overline{E} = E_0(\theta, \phi)\cos(\omega t)\overline{\theta} + E_0(\theta, \phi)\sin(\omega t)\overline{\phi} \tag{1.43}$$

The angle of the total field is now

$$\delta = \tan^{-1}\left[\frac{\sin(\omega t)}{\cos(\omega t)}\right] = \omega t \tag{1.44}$$

which means that the field vector traces out a circle as in Figure 1.5(b). This is called *circular polarization*. If the vector rotates counterclockwise as the wave approaches the observer, the field is *right-hand circularly polarized* (RHCP). Conversely, clockwise rotation indicates *left-hand circular polarization* (LHCP).

The most general case occurs when $|E_\theta(\theta, \phi)| \neq |E_\phi(\theta, \phi)|$ and $\alpha \neq 0$ degrees or 180 degrees. Here the vector traces out an ellipse, which defines elliptical polarization. This is the most general polarization state. The major and minor axes of the ellipse are determined by the relative amplitudes and phases of the two electric field components. The ratio of the major axis to the minor is called the axial ratio. Generally, the axial ratio is used to describe the quality of a circularly polarized wave or antenna and is usually expressed in decibels (for example, 20 log(major axis/minor axis). For a circularly polarized field, the axes are equal, resulting in an axial ratio of 0 dB. A linearly polarized field has only one component; thus, the minor axis is zero, producing an axial ratio of ∞ dB. The smaller the axial ratio, the closer to a circular polarized wave.

Most systems are designed to be either linear or circularly polarized. Polarization is important for several reasons. An antenna extracts the maximum power from an incident wave when the antenna and wave polarizations are identical. No power is received when the polarizations are orthogonal to each other, as for example vertical and horizontal or RHCP and LHCP polarizations. Orthogonal polarizations are used to increase spectrum utilization in systems sharing frequency allocations and perhaps propagation paths. Circularly polarized antennas have no preferred orientation, which makes them useful for systems where the desired orientation changes or is unknown, as in satellite-to-mobile user applications.

The last two antenna parameters are impedance and bandwidth. Antenna impedance is exactly the same as for any other component; that is, it is the ratio voltage across the antenna terminals to the current flowing into the antenna. Impedance can be the most difficult parameter to calculate because it strongly depends upon the accuracy of the model in the feed region of the antenna. The bandwidth has no unique definition for an antenna. Each parameter such as radiation pattern, gain, and impedance may behave differently with frequency. For example, the radiation pattern of small dipole is independent of frequency as long as the dipole is short compared to the wavelength. The impedance, however, varies rapidly with frequency because a small dipole has a high quality factor or Q. In situations such as this, it is usually the highly variable parameter that defines bandwidth. For other cases, the behavior of several parameters defines the bandwidth. It may be required to have the 3-dB beamwidth to be less than 10 degrees, maximum sidelobe level be less than −15 dB, and the return loss be less than 10 dB over a specified bandwidth.

1.7 ORGANIZATION OF THE BOOK

The purpose of this book is to walk the reader through the design process for several types and configurations of microstrip antennas. Along the way computer programs are introduced that either implement the design procedure or provide tools to assist in the design. Successful design requires a good understanding of microstrip antenna operation. Many models have been proposed to analyze microstrip antennas. Several will be addressed in Chapter 2. No detailed mathematics will be presented. Most of the models are based upon a physical viewpoint of the antenna and as such not only provide a means for analysis but also supply insight into the antenna physics. This is the main reason for discussing the models. Having appropriate insight is an invaluable tool for good design. For those readers interested in probing more deeply into the theory and design of microstrip, several excellent texts (for example, [16–20]) are available.

The remaining chapters cover the design of various elements and arrays. Chapter 3 provides an in-depth discussion of the design of basic, single-element antennas. The focus is on rectangular and circular elements and some of their variants. A computer program is included for each element design with a sample case to illustrate program usage. Chapter 4 presents some advanced feeding approaches using feeds with no direct contact with the antenna. These approaches open up new ways of effectively using patch antennas. A few non-microstrip-printed antennas are also discussed. Chapter 5 covers the design of circularly polarized elements with discussions of antennas with two- and single-point feeds. The major limitation of microstrip antennas is a narrow impedance bandwidth. Bandwidth enhancement techniques are presented in Chapter 6. One approach involves the use of an impedance-matching network, another uses stacked patches. Coplanar, parasitically coupled patches are also discussed. When narrower beamwidths and higher gains are required, an array of microstrip elements is needed. Arrays are covered in Chapter 7. Array architectures and design techniques are also presented.

References

[1] Deschamps, G. A., "Microstrip Microwave Antennas," *3rd USAF Symp. Antennas*, University of Illinois, Urbana, IL, 1953.

[2] Gutton, H., and G. Baissinot, "Flat Aerial for Ultra High Frequencies," French Patent no. 703113, 1955.

[3] Howell, J. Q., "Microstrip Antennas," *IEEE Int. Symp. Digest Antennas and Propagation*, Williamsburg, VA, 1972, pp. 177–180.
[4] Weinschel, H. D., "Progress Report on Development of Microstrip Cylindrical Arrays for Sounding Rockets," Physical and Sciences Laboratory, New Mexico State University, Las Cruces, NM, 1973.
[5] Sanford, G. G., "Conformal Microstrip Phased Array for Aircraft Tests with ATS-6," *Proc. Nat. Electron. Conf.*, Vol. 29, Oct. 1974, pp. 252–257.
[6] Munson, R. E., "Conformal Microstrip Antennas and Microstrip Phased Arrays," *IEEE Trans. on Antennas and Propagation*, Vol. 22, No. 1, Jan. 1974, pp. 74–77.
[7] Grieg, D. D., and H. F. Englemann, "Microstrip—A New Transmission Technique for the Kilomegacycle Range," *Proc. IRE*, Vol. 40, 1952, pp. 1644–1650.
[8] Mailoux, R., J. McIlvenna, and N. P. Kernweis, "Microstrip Antennas Technology," *IEEE Trans. on Antennas and Propagation*, Vol. 29, No. 1, Jan. 1981, pp. 25–37.
[9] Ramo, S., J. R. Whinnery, and T. Van Duzer, *Fields and Waves in Communications Electronics*, 2nd ed., New York, NY: John Wiley & Sons, 1984.
[10] Cheng, D. K., *Field and Wave Electromagnetics*, Reading, MA: Addison-Wesley, 1983.
[11] *IEEE Trans. on Antennas and Propagation*, Vol. 17, No. 3, May 1969.
[12] *IEEE Trans. on Antennas and Propagation*, Vol. 22, No. 1, Jan. 1969.
[13] Collin, R. E., *Antennas and Radiowave Propagation*, New York, NY: McGraw-Hill, 1985, pp. 20–25.
[14] Balanis, C. A., *Antenna Theory, Analysis and Design*, New York, NY: Harper & Row, 1982, pp. 27–28.
[15] Collin, R. E., *Antennas and Radiowave Propagation*, New York, NY: 1 McGraw-Hill, 1985, Chap. 5.
[16] Bahl, I. J., and P. Bhartia, *Microstrip Antennas*, Norwood, MA: Artech House, 1980.
[17] James, J. R., P. S. Hall, and C. Wood, *Microstrip Antennas, Theory and Design*, London, UK: Peter Peregrinus, 1981.
[18] James, J. R., and P. S. Hall, eds., *Handbook of Microstrip Antennas*, London, UK: Peter Peregrinus, 1989.
[19] Bhartia, P., K. V. S. Rao, and R. S. Tomar, *Millimeter-Wave Microstrip and Printed Circuit Antennas*, Norwood, MA: Artech House, 1991.
[20] Pozar, D. M., and D. H. Schaubert, eds., *Microstrip Antennas, The Analysis and Design of Microstrip Antennas and Arrays*, New York, NY: IEEE Press, 1995.

Chapter 2
Microstrip Radiator Models

Although microstrip antennas are relatively simple structures, their analysis is quite complicated. The primary complicating factor is the presence of the dielectric substrate with the backing conducting ground plane. A tremendous amount of research has gone into developing analysis approaches and models. Not surprisingly, early models were relatively simple, providing basic design information for certain patch geometries but incapable of accounting for all phenomena. Improved analysis techniques followed that expanded the range of shapes and configurations that could be analyzed but still did not provide rigorous solutions. Rigorous models based on integral equation formulations began appearing in the early 1980s. Recently accurate formulations using differential equations have been reported. Although the rigorous approaches offer the best accuracy, they involve very intensive numerical computations. As a result, the simpler, earlier approaches are still widely used in engineering work. This chapter discusses some of these models, with an emphasis on developing an understanding of the physics of microstrip antennas.

2.1 MICROSTRIP TRANSMISSION LINES

Before beginning a discussion of various models for microstrip antennas, it is worthwhile to cover some aspects of microstrip transmission lines. The microstrip line forms the basis for the antennas, so a knowledge of its behavior helps in understanding the antennas. Figure 2.1 shows the cross section of a microstrip line along with a (very) approximate sketch of the field distribution transverse to the direction of propagation, which is assumed into the paper. The electric field lines go from the microstrip line to the ground plane. Most of the lines are concentrated underneath the microstrip. Some lines originate on charges on the edge and top of the microstrip and therefore partially extend into the free

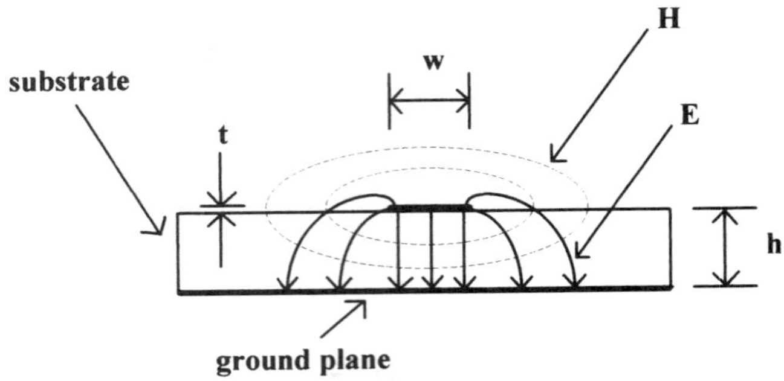

Figure 2.1 Microstrip cross section with field distribution.

space above the substrate. The magnetic field lines encircle the microstrip line and extend above the substrate.

The extension of the fields above the substrate has several implications. The first is that the wave supported by the line is non-TEM. There are longitudinal field components along the direction of propagation that are not shown in the figure. Usually these are significantly smaller than the transverse ones and can, to a first order, be ignored. Second, the presence of field lines in the air reduces the effective dielectric constant seen by waves propagating along the line. If all the fields existed between the line and ground plane, the dielectric constant would be that of the substrate. Instead it is somewhat less than that of the substrate-how much less depends primarily upon the line width, substrate dielectric constant, and, to a lessor extent, the substrate height. Finally the various microstrip parameters such as characteristic impedance and phase velocity become frequency dependent. This phenomenon is called dispersion. Fortunately, the frequency dependency is very small for frequencies below about 8 GHz and for electrically thin substrates.

As with microstrip antennas, a large number of analysis techniques have been applied to microstrip transmission lines. Gupta et al. [1] provide a comprehensive review of this subject. Because of the engineering importance of microstrip, much effort was placed on developing closed-form expressions for the transmission line parameters. The work of Hammerstad [2] and Bahl and Garg [3] achieved wide acceptance as being sufficiently accurate for most design work. The characteristic impedance, Z_0, and effective dielectric constant, ϵ_{re}, are given by

$$Z_0 = \frac{\eta_o}{2\pi\sqrt{\epsilon_{re}}} \ln\left\{\frac{8h}{W_e} + 0.25\frac{W_e}{h}\right\} \quad \text{for } W/h \leq 1 \tag{2.1}$$

$$Z_0 = \frac{\eta_o}{\sqrt{\epsilon_{re}}}\left\{\frac{W_e}{h} + 1.393 + 0.667 \ln\left[\frac{W_e}{h} + 1.444\right]\right\}^{-1} \quad \text{for } W/h \geq 1 \tag{2.2}$$

where

$$\frac{W_e}{h} = \frac{W}{h} + \frac{1.25}{\pi}\frac{t}{h}\left[1 + \ln\left(\frac{4\pi W}{t}\right)\right] \quad \text{for } W/h \leq 1/2\pi \tag{2.3}$$

$$\frac{W_e}{h} = \frac{W}{h} + \frac{1.25}{\pi}\frac{t}{h}\left[1 + \ln\left(\frac{2h}{t}\right)\right] \quad \text{for } W/h \geq 1/2\pi \tag{2.4}$$

and

$$\epsilon_{re} = \frac{\epsilon_r + 1}{2} + \frac{\epsilon_r - 1}{2}F\left(\frac{W}{h}\right) - C \tag{2.5}$$

where

$$F\left(\frac{W}{h}\right) = \left(1 + 12\frac{h}{W}\right)^{-1/2} + 0.04\left(1 - \frac{W}{h}\right)^2 \quad (W/h \leq 1) \tag{2.6}$$

$$F\left(\frac{W}{h}\right) = \left(1 + 12\frac{h}{W}\right)^{-1/2} \quad (W/h \geq 1) \tag{2.7}$$

$$C = \frac{\epsilon_r - 1}{4.6}\frac{\frac{t}{h}}{\sqrt{\frac{W}{h}}} \tag{2.8}$$

Also in the preceding equations, η_0, the free-space wave impedance, equals 377Ω; ϵ_r is the substrate relative dielectric constant; and the other notations are as defined in Figure 2.1. These expressions are frequency independent and valid when dispersion can be ignored.

Above about 8 GHz, dispersion should be included to improve design accuracy. References [1] and [4] present closed-form expressions for the frequency variation. They essentially modify the values obtained from (2.1) to (2.8) to account for the dispersion. In this book, the frequency-dependent characteristic impedance and relative dielectric constant will be denoted by $Z_0(f)$ and $\epsilon_{re}(f)$, respectively.

In many circuits, microstrip line lengths often must be certain fractions of a wavelength. The wavelength on the line is found from

$$\lambda = \frac{\lambda_0}{\sqrt{\epsilon_{re}}} \ or \ \frac{\lambda_0}{\sqrt{\epsilon_{re}(f)}} \tag{2.9}$$

where λ_0 is the free-space wavelength. The phase shift caused by a length of line, l, is

$$\beta = \frac{2\pi}{\lambda} l = \frac{2\pi}{\lambda_0}\sqrt{\epsilon_{re}}l \ or \ \frac{2\pi}{\lambda_0}\sqrt{\epsilon_{re}(f)}l \tag{2.10}$$

Signals traveling down the microstrip experience attenuation caused by the finite conductivity of the conductor and ground plane metalization and losses in the dielectric. This is expressed by the attenuation coefficient. It has two components—one for conductor loss, α_c, and one for dielectric loss, α_d. Expressions for these in decibels per unit length are [1]

$$\alpha_c = 1.38A\frac{R_s}{hZ_0}\frac{(32 - W_e/h)^2}{(32 + W_e/h)^2} \quad \text{for } W/h \leq 1 \tag{2.11}$$

$$\alpha_c = 6.1 \times 10^{-5}A\frac{R_s Z_0 \epsilon_{re}}{h}\left[W_e/h + \frac{0.667W_e/h}{W_e/h + 1.444}\right] \quad \text{for } W/h \geq 1 \tag{2.12}$$

and

$$\alpha_d = 27.3\frac{\epsilon_r}{\epsilon_r - 1}\frac{\epsilon_{re} - 1}{\sqrt{\epsilon_{re}}}\frac{\tan\delta}{\lambda_0} \tag{2.13}$$

where

$$A = 1 + \frac{h}{W_e}\left\{1 + \frac{1.25}{\pi}\ln\left(\frac{2\pi W}{t}\right)\right\} \quad \text{for } W/h \geq 1/2\pi \tag{2.14}$$

$$A = 1 + \frac{h}{W_e}\left\{1 + \frac{1.25}{\pi}\ln\left(\frac{2h}{t}\right)\right\} \quad \text{for } W/h \geq 1/2\pi \tag{2.15}$$

$$R_s = \sqrt{\frac{\pi f \mu_0}{\sigma}} \tag{2.16}$$

where σ is the conductor conductivity in Siemens per meter (S/m). The total attenuation coefficient is $\alpha = \alpha_c + \alpha_d$.

Consider a signal of magnitude V_0 that is input on a microstrip line. After traveling a distance, l, the signal is

$$V = V_0 e^{-\alpha l} e^{-j\beta l} \tag{2.17}$$

which follows from standard transmission line theory. The signal magnitude is reduced by the attenuation (the exponential term containing α) and experiences a phase shift (the exponential term with β). α is the attenuation constant, and β is the phase constant. They are combined to form the quantity $\gamma = \alpha + j\beta$, which is called the propagation constant. Most microstrip lines have negligible loss so that $\gamma \approx j\beta$. γ along with Z_0 completely characterize the line.

Although not explicitly stated earlier, the prior results apply to a microstrip line where only the lowest order, dominant mode (as sketched in Figure 2.1) is present. The grounded dielectric substrate (without the microstrip line) can guide energy as well. These guided waves are called surface waves. Most of the surface wave fields are located within the dielectric. Some field structure exists in the free space above the substrate. These fields decay exponentially as one moves vertically away from the substrate. The lowest order surface wave mode for the substrate has a zero cut-off frequency and therefore can be excited at any frequency. This is the TM_0-mode. The other surface wave modes have nonzero cut-off frequencies. For electrically thin substrates, these higher order modes are usually below cut-off and are not readily excited.

Signals on microstrip circuits can excite a surface wave, usually the TM_0 mode in the substrate. This results in lost power since the surface wave typically propagates away from the microstrip. The surface wave continues to propagate until it meets a discontinuity such as the edge of the substrate or another microstrip line. When the surface wave "strikes" the discontinuity, it may radiate and/or couple energy to the discontinuity. Both of these are undesirable as they usually disrupt circuit or antenna performance. The excitation of the TM_0 mode is minimized by choosing thin substrate heights and low dielectric constants. For a given relative dielectric constant, the height should be chosen such that [5]

$$h \leq \frac{c}{4f\sqrt{\epsilon_r - 1}} \tag{2.18}$$

where c is the speed of light, 3×10^8 m/s, and f is the operating frequency.

The electric field underneath the microstrip line, as shown in Figure 2.1, is essentially uniform across the line. As the frequency or line width increases it is possible to excite a transverse resonant mode on the line [5]. The microstrip line behaves essentially like a resonator with a standing wave excited along its width. The electric field is a maximum at the line edge, goes to zero at the center, and then returns to a maximum at the other edge. Obviously this is undesirable as power becomes lost into this mode. Signal distortion can also occur because the energy gets split into two modes that travel at different velocities. For a given substrate and line width, the cut-off frequency for the transverse mode is given by [5]

$$f_c = \frac{c}{\sqrt{\epsilon_r}[2w + 0.8h]} \tag{2.19}$$

The maximum line width that should be used is such that the cut-off frequency, as given by (2.19), is well above the maximum frequency of interest.

A program that incorporates (2.1) to (2.8) and (2.11) to (2.16) plus corrections for dispersion is described in Appendix A. This program calculates the characteristic impedance and effective dielectric constant for a given substrate, line width, and frequency.

2.2 MICROSTRIP DISCONTINUITIES

Microstrip circuits invariably consist of more than a single straight line. Lines are often bent because of, for example, circuit topology and substrate area limitations. Changes in line width are needed when different characteristic impedances are required. Junctions of three and four microstrip lines are used in various ways. Open-circuited stubs are used for impedance matching and to simulate short circuits. A gap or a notch can be introduced in a line. All of these represent abrupt changes in the microstrip line geometry and are called discontinuities.

Discontinuities alter the electric and magnetic field distributions. This results in energy storage and sometimes radiation at the discontinuity. Usually the physical size of a discontinuity is much smaller than a wavelength. In this case the discontinuity can be represented by an equivalent circuit consisting of lumped elements such as capacitors, inductors, and, if there is radiation, resistors. Changes in the electric field at a discontinuity produce additional capacitance while changes in the magnetic field introduce inductance. In most designs, the effects of discontinuities must be accounted for; otherwise, circuit performance will not be as expected. For example, quarter-wavelength-long open-circuited stubs are often used for impedance matching. The abrupt truncation of the microstrip line at the open circuit results in fringing electric fields that extend beyond the line. The fringing fields store energy and act like a capacitor connected to the end of the line. Because of this stray capacitance, the line is electrically longer than its physical length.

Many techniques have been developed for determining the equivalent circuits for various microstrip discontinuities. At frequencies up to the lower end of the microwave region, a quasi-static analysis usually provides circuit elements accurate enough for most design purposes. Figure 2.2 shows several common discontinuities and their equivalent circuits [6]. Figure 2.2(a) shows the open-circuited microstrip line. Its equivalent circuit is a capacitor to ground, representing stored energy in the electric fields that fringe out beyond the open end. These field lines run from the microstrip line to the ground plane. It is known from transmission line theory that a capacitance can be represented by a section of open-circuited transmission line. Effectively the shunt capacitance makes the open-circuited line electrically longer than its physical length. This is referred to as the open-end length extension. Thus the open end can be represented by either the original line plus a shunt capacitance or a new line that is physically longer. Expressions for this extension are given in [6], while [1] contains expressions for the equivalent capacitance.

The step change in line width is shown in Figure 2.2(b). Both the electric and magnetic fields are altered at the width change. The equivalent circuit therefore contains

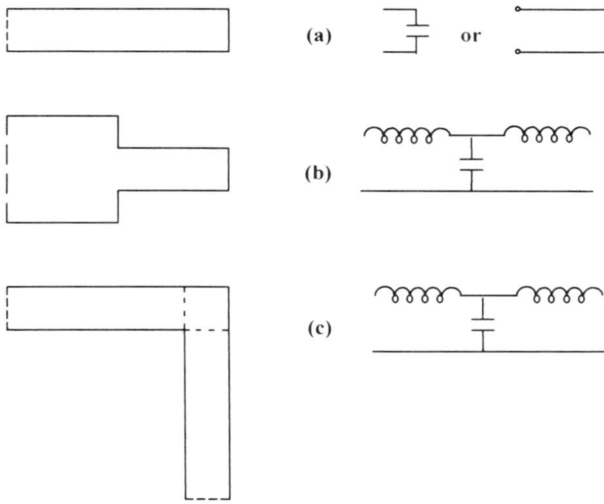

Figure 2.2 Some microstrip discontinuities and their equivalent circuits: (a) open end, (b) step change in width, and (c) microstrip bend.

inductors and capacitors. Expressions for these are given in [1] and [6]. A bend in a microstrip line is also a discontinuity as the bend causes crowding of current around the corner and a creation of excess charge at the outside corner of the bend. The bend shares the same equivalent circuit as the width change but obviously with different expressions for the components [1,6]. The equivalent circuit for the bend represents electrical behavior of the physical area defined by the outside corner and the dashed lines at the corner in the figure. As a result the electrical length of each line forming the bend does not go to the center of the intersection with the other line but rather to the dashed line. These dashed lines are the electrical reference planes for the discontinuity. The discontinuities of Figures 2.2(a,b) have only one reference plane that is located at the physical location of the discontinuity.

Another important discontinuity is the T-junction as given in Figure 2.3. Possible reference planes are shown as dashed lines at the junction in the figure. The transformers represent the changes in impedance levels caused by the width changes. The shunt susceptance plus the movement of the reference planes out from the junction center account for stored energy.

The bend occurs often enough in microstrip circuits to make it awkward to have to continually account for the discontinuity reactances it introduces. It has been found that the reactances can be minimized by chamfering the corner as in Figure 2.4. The amount that has to be cut off the corner is [6]

$$\frac{x}{d} = 0.52 + 0.65e^{-1.35\frac{w}{h}} \tag{2.20}$$

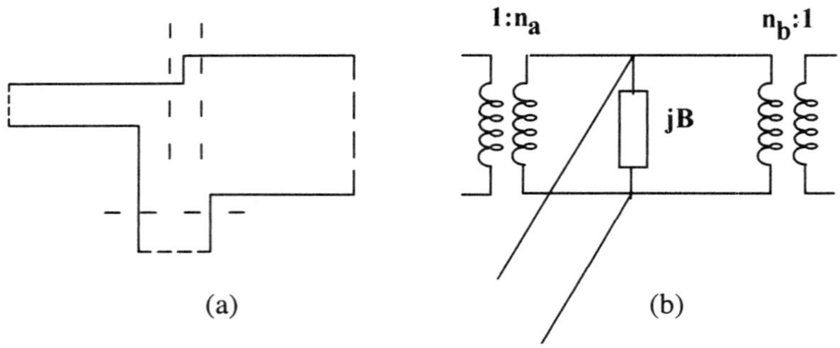

Figure 2.3 (a) Microstrip T-junction and (b) equivalent circuit.

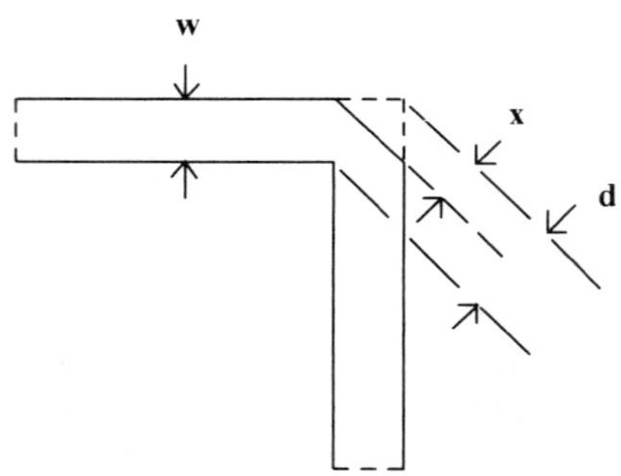

Figure 2.4 A chamfered microstrip bend to minimize the discontinuity reactance.

A bend with this chamfer creates virtually no mismatch and hence looks essentially transparent to any signal. The chamfered bend is electrically shorter than the physical distance along the centerline of the bend. This reduction from the centerline length is [6]

$$\frac{\Delta b}{D} = 0.16[2 - (f/f_p)^2] \tag{2.21}$$

where $D = (\eta_0/\sqrt{\epsilon_r})h/Z_0$ and f_p (GHz) $= 0.4Z_0/h$ (mm).

Simple programs for calculating the equivalent circuit parameters are described in Appendix A. The discontinuities covered are the bend, open-end effect, gap, notch, step width change, and T-junction. Also presented is a program that calculates the optimum miter and the change in electrical length of a chamfered bend.

2.3 MICROSTRIP PATCH TRANSMISSION LINE MODEL

In Chapter 1 it was mentioned that microstrip antennas are essentially suitably shaped discontinuities that are designed to radiate. A rectangle with a length of about 0.5 of a wavelength in the dielectric and a width of from 0.5 to 2.0 times the width is one such example. Most of the earliest microstrip antennas were rectangular in shape. The earliest and simplest model of a rectangular microstrip radiator is based upon treating it as a microstrip transmission line that radiates [7–9] from two of its ends. Even today, the transmission line model is still very useful for design work [10]. It is based on a somewhat crude but reasonably accurate physical picture of radiation from the antenna. As such, the model provides significant insight into antenna operation. Although it has been extended to cover other shapes, the primary application of the transmission line model has been for rectangular patches.

Consider a rectangular microstrip antenna fed by a microstrip line as presented in Figure 2.5. When the fields on the microstrip line encounter the abrupt change in width at the input to the patch, they spread out (only electric fields are shown in the figure). This creates fringing fields at this edge, as indicated in the figure. After this transition, the patch looks like another microstrip line. The fields propagate down this transmission line until the other edge is reached. Here the abrupt ending of the line again creates fringing fields as was discussed earlier for the open-end discontinuity. The fringing fields store energy. The edges appear as capacitors to ground since the changes in the electric field distribution are greater than that for the magnetic field. Because the patch is much wider than a typical microstrip line, the fringing fields also radiate, which is represented by a conductance in shunt with the edge capacitance, which accounts for power lost due to radiation.

An equivalent circuit for the microstrip patch can now be constructed from this physical interpretation. Each edge is a parallel combination of a shunt capacitor and conductance. The edges are connected by a length of transmission line resulting in the equivalent circuit of Figure 2.6. The input impedance for this circuit is the parallel

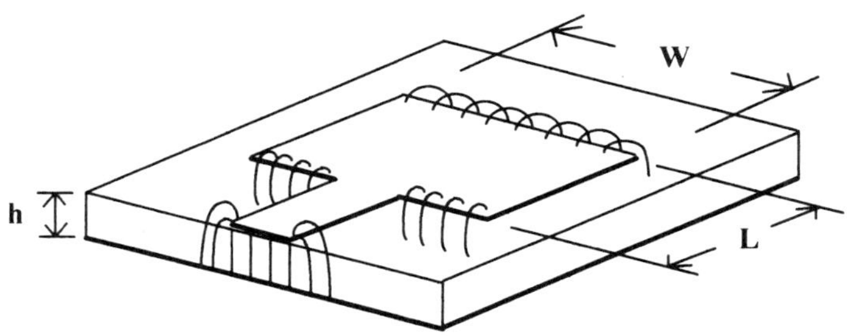

Figure 2.5 Microstrip antenna radiation mechanism as described by the transmission line model.

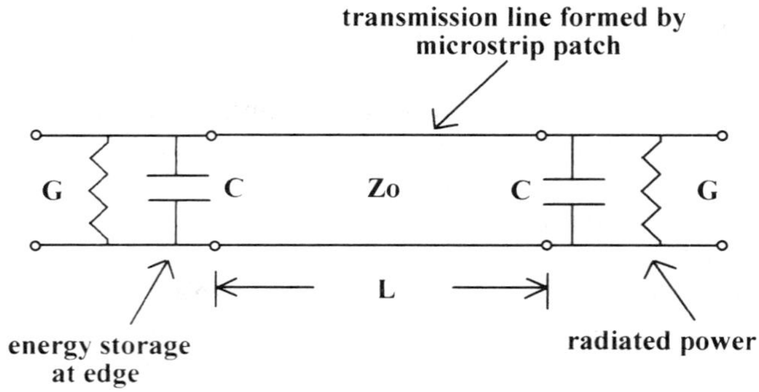

Figure 2.6 Equivalent circuit for a microstrip patch antenna.

combination of the input-edge shunt capacitance and conductance with the admittance of the output edge rotated through a section of transmission line of impedance Z_0 and length L. Working with admittances, this can be expressed as

$$Y_{in} = G + jB + Y_0 \frac{G + j[B + Y_0 \tan(\beta L)]}{Y_0 + j[G + jB \tan(\beta L)]} \tag{2.22}$$

where $Y_0 = 1/Z_0$, $B = \omega C$, L is the patch length, and β is the phase constant of the microstrip line. When L is approximately 0.5 a wavelength, the susceptance of the last term in (2.22) is the negative of that of the input edge. The input admittance and therefore the input impedance become entirely real. This defines the operating frequency of the patch. The patch is said to be resonant at this frequency.

When the patch is 0.5 a wavelength long, the fringing fields of the output edge are out of phase with respect to those at the input edge. Assume the input-edge fields are positive; that is, they "point" upward from the ground plane to the patch . The output-edge fields will then point downward toward the ground plane. When looking directly down on the patch, the fringing fields point in the same direction as in Figure 2.7. The radiation from these fields add up to produce a far-field pattern with a maximum broadside to the patch.

Now that the equivalent circuit has been established, the next step is to determine values for the various circuit elements. The characteristics of the transmission line created by the patch can be found using (2.1) to (2.8) and (2.10) to (2.16) plus any corrections for dispersion if needed. These equations calculate the characteristic impedance and propagation constant for the patch line. Once ϵ_{re} (or $\epsilon_{re}(f)$) is known, the approximate patch length is $0.5\lambda_0/\sqrt{\epsilon_{re}}$.

The edge capacitance is found from the end-effect extension. Care has to be taken because many expressions for the end effect are only valid for the relatively narrow line widths typically used for microstrip lines. One formulation has been reported [11] that is

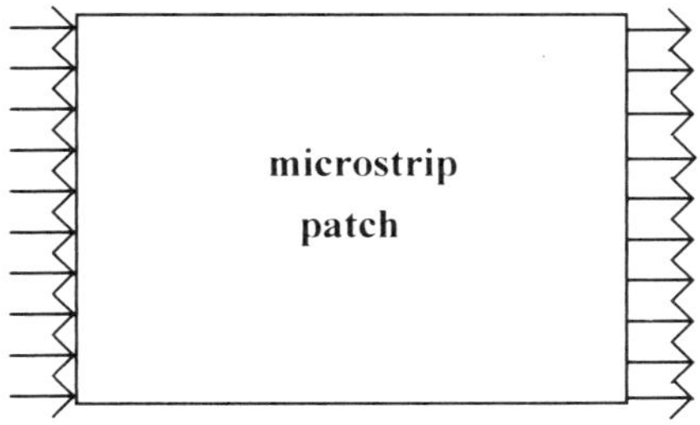

Figure 2.7 Fringing fields at patch edges.

valid for $0.01 \leq W/h \leq 100$ and $\epsilon_r \leq 128$. It is based on curve-fitting results from a very accurate full-wave analysis of the open end. The end-effect extension, Δl, of an open end beyond its physical end is

$$\Delta l = \frac{\zeta_1 \zeta_3 \zeta_5}{\zeta_4} h \tag{2.23}$$

where

$$\zeta_1 = 0.434907 \frac{\epsilon_{ere}^{0.81} + 0.26}{\epsilon_{ere}^{0.81} - 0.189} \frac{(w/h)^{0.8544} + 0.236}{(w/h)^{0.8544} + 0.87}$$

$$\zeta_2 = 1 + \frac{(w/h)^{0.371}}{2.358\epsilon_r + 1}$$

$$\zeta_3 = 1 + \frac{0.5274 \tan^{-1}[0.084(w/h)^{1.9413/\zeta_2}]}{\epsilon_{re}^{0.9236}}$$

$$\zeta_4 = 1 + 0.5274 \tan^{-1}[0.067(w/h)^{1.456}]\{6 - 5e^{0.036(1-\epsilon_r)}\}$$

$$\zeta_5 = 1 - 0.218e^{-7.5w/h}$$

The end-effect extension can be related to the edge capacitance using transmission line theory. The end effect is essentially a very short section of open-circuited transmission line. It therefore presents a certain impedance to the patch edge. Since the end effect is small ($\ll \lambda/4$), this impedance will be a negative reactance just as a capacitor. Setting the impedance equal to that of the capacitor's, we can solve for the capacitance

$$C = \frac{\tan(\beta \Delta l)}{\omega Z_0} \tag{2.24}$$

where $\omega = 2\pi f$. Equation (2.24) links the end-effect extension to an equivalent capacitance.

The edge conductance represents power radiated away by the microstrip. This is found from the radiation from the edge that has two components—one the normal space wave radiation as with any antenna and the other caused by surface wave excitation. For substrate heights that are very small compared to the wavelength ($h < 0.02\lambda_0$), surface wave radiation is negligible and can be ignored. When the substrate is thin, the effect of the dielectric on the radiation pattern can be ignored to a first order as well. The patch and substrate can then be removed and replaced by two slots located where the edges originally were on the infinite ground plane. The slot length and width are approximately the patch width, W, and the substrate height, h.

Since the electric field underneath the patch is essentially uniform across the patch width, the slot has a similar uniform field. It is relatively straightforward to calculate the radiation pattern for the slot. The geometry for the slot is shown in Figure 2.8.

Assuming that the patch lies in the (x-z)-plane, the H-plane pattern for the slot is in the (y-z)-plane that lies along the patch width [8]

$$E_\phi = V_0 \frac{\sin[(k_0 W/2)\cos(\theta)]}{(k_0 W/2)\cos(\theta)} \tag{2.25}$$

while the E-plane pattern lies in the (x-y)-plane (along the patch length) and is

$$E_\theta = V_0 \frac{\sin[(k_0 h/2)\cos(\phi)]}{(k_0 h/2)\cos(\phi)} \tag{2.26}$$

where V_0 is the voltage across the slot and $k_0 = 2\pi/\lambda_0$. In (2.25) θ is measured from the z-axis. For (2.26) ϕ is measured from the x-axis. Since $h << \lambda_0$, (2.26) is basically unity

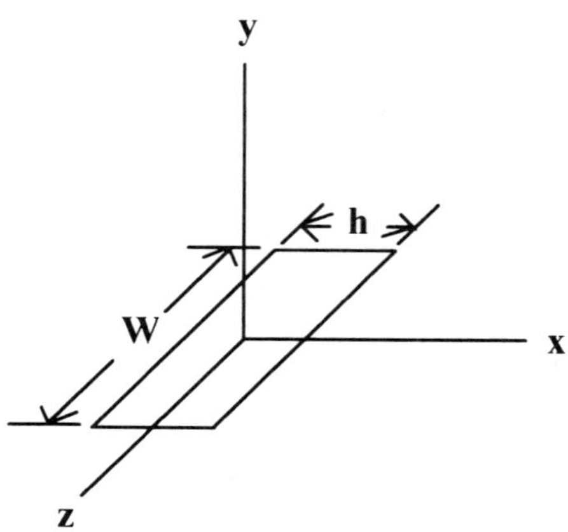

Figure 2.8 Coordinate system for radiating edge equivalent slot in the ground plane.

for all angles. The power radiated by the slot is found by integrating the Poynting vector over the hemisphere above the ground plane. Radiated power is power that is lost in the sense that it leaves the antenna. Obviously in this case it is a desirable effect. When a voltage is placed across a conductance, current flows and power is also lost. The power lost due to radiation can be equated to power lost in a conductance. An equivalent conductance that dissipates the same power that was radiated can be found by dividing the integral of the Poynting vector, which is a power, by $V_0^2/2$. The resulting expression for the edge conductance is

$$G = \frac{1}{\pi}\sqrt{\frac{\epsilon}{\mu}}\int \frac{\sin^2[(\pi W/\lambda_0)\cos(\theta)]}{\cos^2(\theta)}\sin^3(\theta)\, d\theta \tag{2.27}$$

In [10], the following expression is reduced to the closed form for easy evaluation

$$G = \frac{1}{\pi^2\eta_0}\left\{\left(wSi(w) + \frac{\sin w}{w} + \cos w - 2\right)\left(1 - \frac{s^2}{24}\right) + \frac{s^2}{12}\left(\frac{1}{3} + \frac{\cos w}{w^2} - \frac{\sin w}{w^3}\right)\right\} \tag{2.28}$$

where $w = k_0 W$, $Si(w)$ is the sine integral, and $s = k_0 \Delta l$.

As presented so far, the transmission line model does not account for all phenomena associated with the patch. At each edge there can be some surface wave excitation. If the substrate is thin ($h < 0.02\lambda_0$), the amount of power lost to surface waves is much less than that due to radiation. This limits the validity of the transmission line model to thin substrates. It is not easy to include surface-wave effects because the calculation of surface-wave power is quite complicated. Another factor not accounted for is electromagnetic coupling between the radiating edges. The fields from one edge extend out into the region occupied by the other edge and vice versa. The edges are thus coupled externally by their respective fields (they are coupled internally by the patch transmission line).

Electromagnetic coupling between antennas is a commonly studied problem. The reciprocity theorem and the known fields from each antenna are used to derive an integral expression for the mutual coupling. If the edges are excited by equal voltages (which they normally are), the mutual admittance, Y_m, between them is given by [12]

$$Y_m = \frac{1}{|V_0^2|}\int \overline{E_1} \times \overline{H_2^*} \cdot \overline{dS} \tag{2.29}$$

where $\overline{E_1}$ is the electric field of one edge, $\overline{H_2}$ is the magnetic field of the other, $\overline{dS}$ is a vector normal to a large hemisphere surrounding the patch, and the asterisk denotes the complex conjugate. Again (2.29) has been reduced to closed form [10]. The real part of Y_m is denoted by G_m and is

$$G_m = \left\{ J_0(l) + \frac{s^2}{24 - s^2} J_2(l) \right\} G \tag{2.30}$$

where $J_0(l)$ and $J_2(l)$ are the Bessel functions of the first kind, order 0 and 2, respectively, $l = k_0 L$, and $s = k_0 \Delta l$. B_m is the imaginary part of Y_m and is denoted

$$B_m = \frac{\pi}{2} \frac{Y_0(l) + [s^2/(24 - s^2)]Y_2(l)}{\ln(s/2) + 0.577216 - 1.5 + (s^2/12)/(24 - s^2)} [1 - e^{-0.21w}]\omega C \tag{2.31}$$

where $Y_0(l)$ and $Y_2(l)$ are the Bessel functions of the second kind, order 0 and 2, respectively.

The electromagnetic coupling is incorporated into the transmission line model by adding two dependent current sources to the equivalent circuit. Figure 2.9 shows the modified circuit. At resonance, the input resistance for a patch fed at one edge is usually quite high, being on the order of 100Ω to 300Ω. This necessitates the use of a matching circuit, like a quarterwave transformer, since the normal impedance level for microstrip circuits is 50Ω. The input resistance can be reduced by moving the feed inward toward the center of the patch. This is discussed in Chapter 3. The input resistance varies approximately as a cosine squared function of the inset distance, having a maximum at the edge and going to zero at the center. An inset feed is indicated in Figure 2.9 where the feed has been inset a distance L_1. Reference [10] contains a generalized expression for the input impedance, including an inset feed and attenuation ($\alpha \neq 0$) in the patch transmission line. This expression is used in the computer model presented in Chapter 3.

Earlier it was stated that the patch length is about a half-wavelength long, that is, $L \approx 0.5\lambda_0/\sqrt{\epsilon_r}$. Because of the end-effect extension, the patch electrical length is longer than its physical length. The actual patch design length is more accurately given by $L_p \approx L - 2\Delta l$, which is the physical length minus twice the end-effect extension. The above approximation is still reasonably accurate especially for initial sizing of the antenna.

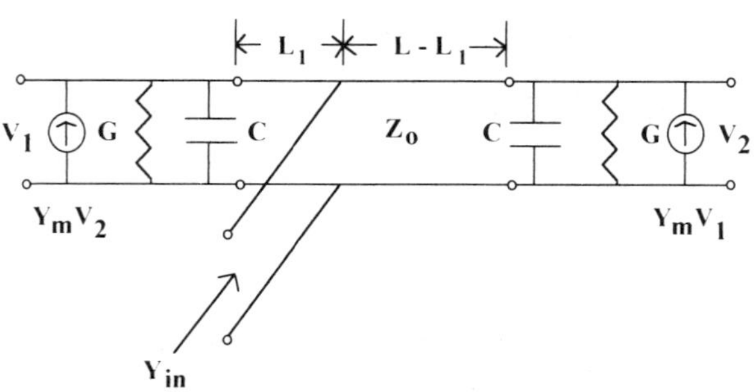

Figure 2.9 Transmission line equivalent circuit including edge mutual coupling and inset feed.

(The computer programs of Chapter 3 provide a more accurate estimate of length.) It was also mentioned earlier that the microstrip line width should not be so wide as to excite the transverse resonance mode. Microstrip patches are quite wide and may violate the criteria given by (2.19). Placing the feed at the center of the patch width forces the electric field to be a maximum there. This suppresses the excitation of the transverse mode as long as the width does not exceed a wavelength in the dielectric.

2.4 MICROSTRIP PATCH RADIATION PATTERNS

The radiation from the patch is linearly polarized with the electric field lying in the same direction as the patch length. The patch has two radiating edges, and they form a two-element array. Expressions for the E- and H-plane patterns of each edge were given in (2.25) and (2.26). The two-element array modifies the patch pattern in the E-plane to [9]

$$E_\theta = V_0 \frac{\sin[(k_0 h/2)\cos(\phi)]}{(k_0 h/2)\cos(\phi)} \cos\left[\frac{k_0 L}{2}\cos(\phi)\right] \tag{2.32}$$

The last factor is due to the two slots that are separated by L. The H-plane pattern for the patch is still given by (2.25). Equations (2.32) and (2.25) do not account for the presence of the dielectric substrate. They are reasonably accurate except in the E-plane when the angle approaches the substrate (for example, $\phi \to 0$ degrees or 180 degrees).

The substrate can be included in the patterns using a technique based on the reciprocity theorem [13]. Consider two infinitesimal dipoles, one located on the surface of the dielectric and the other at a very large distance away. Since the second dipole is very far away, its radiated field looks like a plane wave when it reaches the dielectric surface. The total field at the surface due to the distant dipole is the sum of the direct field plus the field that enters the dielectric and reflects off the ground plane. This total field is found in a relatively straightforward manner using wave matrices [14]. By applying the reciprocity theorem and the total field, the field at the distant dipole as radiated by the one on the dielectric surface can be found. Superposition is then applied using the actual current distribution on the patch to determine its radiation pattern.

The results of this analysis are the generation of two factors that modify (2.25) and (2.32) to include the effect of the substrate on the radiation pattern. The E-plane and H-plane patterns become

$$E_\theta = V_0 \frac{\sin[(k_0 h/2)\cos(\phi)]}{(k_0 h/2)\cos(\phi)} \cos\left[\frac{k_0 L}{2}\cos(\phi)\right] F(\phi) \tag{2.33}$$

$$E_\phi = V_0 \frac{\sin[(k_0 W/2)\cos(\theta)]}{(k_0 W/2)\cos(\theta)} G(\theta) \tag{2.34}$$

where

$$F(\phi) = \frac{2\tan(\beta_1 h)\cos(\phi)}{\tan(\beta_1 h) - j\epsilon_r\cos(\phi)/n_1(\phi)} \tag{2.35}$$

$$G(\theta) = \frac{2\tan(\beta_1 h)}{\tan(\beta_1 h) - j\eta_1(\theta)\sec(\theta)/\mu_r} \tag{2.36}$$

with $\beta_1 = k_0 n_1(\varphi)$, $n_1(\varphi) = \sqrt{\eta_1^2 - \sin^2(\phi)}$, $\eta_1 = \sqrt{\epsilon_r\mu_r}$, and $\varphi =$ either θ or ϕ. Equations (2.33) and (2.34) are contained in one of the programs discussed in Chapter 3 for the calculation of the patterns for a rectangular patch.

2.5 MICROSTRIP PATCH CAVITY MODEL

Resonator circuits are commonly used in RF and microwave design in circuits such as filters, amplifiers, and oscillators. Below a few gigahertz or so, resonators take the form of lumped element circuits consisting of inductors and capacitors. At higher frequencies, sections of transmission line serve as resonators. Closed sections of waveguide, called cavities, are also used for resonators with rectangular and cylindrical shapes being the most widely popular. These cavities are in the form of a closed box with, for example, the walls and ends all metal. Enclosed cavities avoid power loss due to radiation. Like all resonators, electric and magnetic energy is stored in the cavity. Power is dissipated in the walls because of the finite conductivity of the metal and in any dielectrics because of loss mechanisms in them. The cavity Q is defined in the usual way as

$$Q = \frac{2\pi f W_e}{P_d} \tag{2.37}$$

where W_e is the average stored energy per cycle, and P_d is the dissipated power.

The electric field goes to zero at the metal walls of the cavity. Although it may not be rigorously defined for the cavity, the voltage on the wall is also zero since it is the integral of the electric field. The walls therefore act like a short circuit. A wall with an infinite conductivity is called a perfect electric conductor. The conductivity of many metals is high enough that they can be considered as perfect electric conductors for most practical purposes. It is possible to postulate a perfect magnetic conductor on which the magnetic field goes to zero. At a perfect magnetic conductor, the current density and, therefore the current, go to zero since they are related to the magnetic field. The conductor acts like an open circuit. In addition to using electric conductors, resonators can be constructed using magnetic conductors or combinations of electric and magnetic ones.

Now consider the fields associated with a rectangular microstrip patch [15]. Let the patch substrate lie in the $(x\text{-}y)$-plane. The feed excites electromagnetic fields in the region between the lower side of the patch and the ground plane. When the substrate thickness is much less than a wavelength, the electric field has only a z-directed component that goes from the patch to the ground plane. The magnetic field only has transverse x- and y-directed components between the patch and ground. These fields do not vary with z because the substrate is thin.

A charge distribution is established on the underside of the patch and the ground plane (plus a current flow) [15]. At a particular instant in time, there is an accumulation of positive charge on the underside of the patch and negative charge on the ground plane. The attractive force between these charges tends to keep a large percentage of the charge between the two surfaces. The repulsive force between positive charges on the patch pushes some of these charges around the edge onto the top of the patch. For very thin substrates, the amount of charge pushed onto the top is very small. With little charge flow around the edge, it is reasonable to assume that the current goes to zero there. Perfect magnetic conductors can therefore be introduced at all four walls around the patch without greatly disturbing the field distribution.

A rectangular patch can thus be viewed as a cavity resonator with perfect magnetic conductors for walls and perfect electric conductors at the top and bottom. This allows one to solve the wave equation (as derived from Maxwell's equations) for the electromagnetic field distribution inside the patch. The distribution is only a function of the patch geometry [15]. For geometries in which the wave equation is separable, like rectangular and circular cavities, the solution is straightforward. The amplitudes of the field components, however, depend upon details such as the materials inside the cavity, losses, and the feed structure. Both the distribution and amplitudes are needed to find parameters of the patch such as impedance. The antenna has losses, most of which (hopefully) are associated with radiation. These losses result in the impedance having a real part. To complete the analogy between the cavity and the patch, an approach for incorporating the antenna losses into the cavity is needed.

Consider a cavity with perfectly conducting walls but containing a dielectric with loss. Its Q is given by [15]

$$Q = \frac{1}{\tan \delta} \tag{2.38}$$

where $\tan \delta$ is the loss tangent of the dielectric. When loss is present, the cavity impedance contains a real part representing power loss. The mathematical function for the impedance contains complex poles. The imaginary part of the poles is associated with power dissipation. It can be shown [15] that near the resonant frequency, f_r, the Q is related the impedance function pole by

$$Q = \frac{1}{2} \frac{\omega_r}{\omega_i} \tag{2.39}$$

where ω_r is the real part of the pole and ω_i is the imaginary part. Equations (2.38) and (2.39) relate the losses to the impedance function poles for the cavity. The antenna has complex poles because of radiation and other losses. If the losses in the dielectric-filled cavity are adjusted to match those of the antenna, then the cavity will have the same complex poles as the antenna. It will therefore have the same impedance. From (2.38), the poles will be equal if the dielectric loss tangent is chosen to be the reciprocal of the antenna Q [15]. This completes the analogy between the cavity and the patch and is the basis for the cavity model [15,16].

The cavity model will now be applied to a rectangular patch and its feed [15–17]. Let the patch lie in the (x-y)-plane where the patch width is along the x-axis. Starting with Maxwell's equations and assuming a thin substrate, a wave equation can be derived for the z-component of the electric field underneath the patch. A solution can be found subject to the boundary conditions that the tangential magnetic field is zero on each wall and the tangential electric field is zero on the patch and ground plane. The solution is [18]

$$E_z(x, y) = jI_0\sqrt{\frac{\mu_0}{\epsilon}}k\sum_{m=0}^{\infty}\sum_{n=0}^{\infty}\frac{\psi_{mn}(x, y)\psi_{mn}(x_0, y_0)}{k^2 - k_{mn}^2}G_{mn} \tag{2.40}$$

where I_0 is the magnitude of the feed current, $\epsilon = \epsilon_0\epsilon_r$, $k = \omega\sqrt{\mu_0\epsilon}$, (x_0, y_0) are the coordinates of the feed position, $k_{mn}^2 = k_m^2 + k_n^2$, G_{mn} is an amplitude coefficient related to the feed geometry, and

$$\psi_{mn} = \frac{\chi_{mn}}{\sqrt{WL}}\cos(k_n x)\cos(k_m y) \tag{2.41}$$

In (2.41)

$$\chi_{mn} = \begin{cases} 1, & m = 0 \text{ and } n = 0, \\ \sqrt{2}, & m \neq 0 \text{ or } n = 0, \\ 2, & m \neq 0 \text{ and } n \neq 0, \end{cases}$$

with $k_n = n\pi/W$ and $k_m = m\pi/L$. Each term in (2.40) is a mode for the cavity. Near the patch resonant frequency only one mode is dominant, namely, the $m = 1$, $n = 0$ mode. This is the mode associated with the radiation from (and some energy storage in) the patch. The remaining modes in the summation of (2.40) represent the interaction between the feed and the patch plus additional energy storage. The other field components under the patch can be found from (2.40).

The next step in applying the cavity model is to find the quality factor for the antenna. The major contributor to the patch Q is radiation. The radiation from the patch is found in a manner similar to that used in the transmission line model. The patch is assumed to radiate from its edges. In this case, all the edges are considered since the patch fields are known everywhere inside the patch. The electric field at each edge is found by evaluating (2.40) at the patch boundary. These fields serve as sources for the radiation and are used to find the far-field patterns. Integrating the far-field Poynting vector over a hemisphere surrounding the antenna determines the power radiated by the antenna. If the substrate is very thin and has a low dielectric constant, it is also possible to obtain an approximate value for the surface wave power [15] using the fields at the edges.

After finding the radiated power, the stored energy must be computed. With the fields underneath the patch known from (2.40), the stored energy is determined by

$$W_e = \frac{1}{2}\epsilon_0\epsilon_r \iiint |E_z(x, y)|^2 \, dV \tag{2.42}$$

where the integral is over the volume occupied by the patch. The radiation Q is then

$$Q_r = \frac{2\pi f W_e}{P_{rad}} \tag{2.43}$$

where P_{rad} is the radiated power. If the surface wave power is calculated, its Q is

$$Q_{sur} = \frac{2\pi f W_e}{P_{sur}} \tag{2.44}$$

The other contributors to the patch Q are losses in the conductors and dielectric. It turns out that these losses are relatively independent of the patch shape if the substrate is thin. The Q factors for these are [18]

$$Q_c = h\sqrt{\pi f \mu_0 \sigma} \tag{2.45}$$

and

$$Q_d = \frac{1}{\tan \delta} \tag{2.46}$$

where σ is the conductivity of the patch metal and $\tan \delta$ is the dielectric loss tangent. The total patch Q then becomes

$$\frac{1}{Q} = \frac{1}{Q_r} + \frac{1}{Q_{sur}} + \frac{1}{Q_c} + \frac{1}{Q_d} \tag{2.47}$$

For most microstrip antennas, the radiation term dominates. Typical values for some of the factors in (2.47) are $Q_r \approx 70$, $Q_c \approx 1000$, and $Q_d \approx 400$. The overall patch Q is usually about 55 to 60.

To complete the cavity model, the patch losses must be incorporated. When the cavity dielectric has loss, its relative dielectric constant is complex and is given by $\epsilon_r(1 - j \tan \delta)$ (see (1.16) to (1.18)). With loss, (2.40) still holds, but with ϵ_r replaced with $\epsilon_r(1 - j \tan \delta)$. As shown by (2.38), if the loss tangent is made the reciprocal of the patch Q (2.47), the cavity will have the same fields as the patch. Thus to complete the model, a complex relative dielectric constant is used with $\tan \delta$ set equal to $1/Q$. Any quantity calculated with (2.40) will then be applicable to the patch.

A quantity of major interest to the designer is the patch input impedance. This is the ratio of the voltage across the feed to the current on the feed. One of the previous assumptions is that the fields are independent of z (the direction between the patch and the ground plane). Voltage is the line integral of the electric field. The field at the feed is $E_z(x_0, y_0)$, and since it is not a function of z, the voltage at the feed is $V_f = -hE_z(x_0, y_0)$. After some algebraic manipulation, the input impedance becomes

$$Z_{in} = \frac{V_f}{I_0} = -j\omega\mu_0 h \sum_m^{\infty} \sum_n^{\infty} \frac{\psi_{mn}^2(x_0, y_0)}{k^2 - k_{mn}^2} G_{mn} \tag{2.48}$$

where $k^2 = \epsilon_r[1 - j(1/Q)](\omega^2 \mu_0 \epsilon_0)$. The accuracy of this expression depends quite strongly on the representation of the feed. So far nothing has been said about the G_{mn} terms. They are obtained from an integral that expresses the interaction between the feed and modes of the cavity. The two most common types of feeds are a microstrip line and a coaxial probe. Both are described in Chapter 3. For small feeds with a constant current (or an equivalent current) an approximate expression for G_{mn} is [18]

$$G_{mn} = \frac{\sin(n\pi d_x/2W)}{n\pi d_x/2W} \cdot \frac{\sin(m\pi d_y/2L)}{m\pi d_y/2L} \tag{2.49}$$

where the feed occupies a cross-sectional area that is d_x wide and d_y long. For a microstrip feed, $d_y \approx 0$ and d_x is approximately the width of the line [18]. With a probe feed, $d_y = d_x$ and the product $d_y d_x$ is set equal to the cross-sectional area of the probe [18].

As a final note on the cavity model, when the magnetic walls are placed around the patch, all the fields become confined to the region under the patch. In effect, the fringing fields around the patch are being ignored. Since little current flows on the top surface of the patch, this is a reasonable approximation. The accuracy of the cavity model is enhanced, however, if some attempt is made to include the fringing fields. At both ends, the fields can be accounted for by lengthening the patch using the end-effect extension. This is done by replacing L with $L + 2\Delta l$ in the above equations. The fringing fields on the sides are included by using an effective width, $W_{eff}(f)$

$$W_{\mathrm{eff}}(f) = W + \frac{W_e(0) - W}{1 + f/f_g} \tag{2.50}$$

where f = frequency, $f_g = c/(2W\sqrt{\epsilon_r})$, $c = 3 \times 10^8$ m/S, and

$$W_e(0) = \frac{120\pi h}{Z_0\sqrt{\epsilon_{re}}} \tag{2.51}$$

In (2.51), Z_0 and ϵ_{re} are calculated using (2.1) or (2.2) and (2.5). Equation (2.50) contains the frequency dependency of the effective width, which like other dispersion effects is small for frequencies below about 8 GHz if the substrate is thin. In this situation, (2.51) may be used instead.

2.6 INTEGRAL AND DIFFERENTIAL EQUATION MODELS

The dramatic increase of computer capability in recent years has fostered a tremendous effort in developing numerical techniques for solving electromagnetic problems. In the past, rigorous solutions were restricted to cases in which the geometry or other factors produced equations that could be solved by analytical techniques. For some problems, asymptotic approaches were used. These maintained much of the accuracy of the rigorous formulation while easing the computational problem. In many instances, however, the resulting equations were too difficult to solve with the available techniques. When significant computer power became available, the situation changed. A rigorous solution could be formulated and then solved numerically. The accuracy of the solution essentially becomes a function of the amount of computational effort one is willing to pursue.

Several of these rigorous/numerical solution approaches have been applied to the analysis of microstrip antennas. Although they offer the most accurate results and can be applied to arbitrary structures, the rigorous solutions are quite complicated and beyond the scope of this book. This section will briefly outline some of the approaches. References are given for those readers who wish to pursue these approaches in more detail.

The first and most widely used rigorous approach formulates an integral equation (or equations) appropriate for the problem and then applies a numerical technique called the *moment method* to obtain the solution [19–21]. This technique uses two important concepts of electromagnetics, namely Green's functions and superposition. In an electromagnetic problem, a Green's function is analogous to the impulse response of a circuit. For the given configuration of material bodies (for example, dielectrics and ground planes) and boundaries (for example, walls and infinite free space), Green's function determines the fields at any point due to an infinitesimal uniform current element at a particular location. The actual antenna and circuit occupy a finite volume and may have a current density distribution that varies with position. The fields from the actual structure are found by forming the vector product of the actual current density distribution with Green's function (usually a vector and dyadic) and integrating over the volume occupied by the antenna. This is essentially an application of superposition.

To apply the technique to a microstrip antenna, the geometry and material parameters such as patch shape, feed configuration, number of dielectric layers, and layer characteristics (for example, height ϵ_r and tan δ) are defined. Green's function is found (not a trivial exercise) for this configuration by solving Maxwell's equations. A model is chosen for the feed that serves as the "source" for the problem. Usually it is selected to be a known impressed current density, that is the current established by a generator connected to the feed. This current produces fields that induce a current density on the patch. The patch current density, in turn, produces its own fields. The fields from the impressed feed and patch currents are found using Green's function and superposition as indicated earlier.

Both of these sets of fields, when combined, must satisfy Maxwell's equations and the boundary conditions of the problem. A property of the Green's function is that the fields derived from it automatically satisfy Maxwell's equations and the boundary conditions at any dielectric layer interface and any ground plane. The only condition not satisfied is that the tangential component of the electric field must go to zero on the surface of the patch. This is because the Green's function was derived without the patch present. Mathematically this boundary condition can be expressed as

$$\bar{n} \times [\bar{E}^i(r) + \bar{E}(r)] = 0 \tag{2.52}$$

where $\bar{n}$ is the unit vector of the normal to the surface of the patch, $\bar{E}^i(r)$ is the impressed or source field, and $\bar{E}(r)$ is the field due to patch currents excited by the source. Equation (2.52) is enforced over the area occupied by the antenna.

As indicated above, the fields are found from the current density and the Green's function using superposition. For the patch fields

$$\bar{E}(r) = \iint \bar{\bar{G}}(r, r') \cdot \bar{J}(r')\, dr' \tag{2.53}$$

where $\bar{\bar{G}}(r, r')$ is the Green's function and $\bar{J}(r')$ is the current density on the patch. The integration in (2.53) is over the surface area of the antenna. A similar expression holds for the impressed fields if the source is specified as a current density. The current density on the patch is unknown (the source current is known). When (2.53) is substituted into (2.52), an integral equation is established for the patch current. The solution to this equation is extremely accurate since it satisfies Maxwell's equation and all the boundary conditions. It includes all radiation and surface-wave effects.

Because of the complexity of Green's function and other factors, (2.52) cannot be solved analytically. A numerical solution, usually the method of moments, is used. The unknown patch current is expressed as series of functions with unknown amplitude coefficients as

$$\bar{J}(r') = \sum_{i=1}^{N} \alpha_i f_i(r')\bar{u} \tag{2.54}$$

where the α_i are the unknown amplitude coefficients, the $f_i(r')$ are the expansion functions, and $\bar{u}$ is a unit vector having the same direction as $\bar{J}(r')$. Usually the expansion functions are simple functions like a step, impulse, or triangle that exist only over a small section of the patch. In other words, the patch current distribution is broken up into subsections with an expansion function and coefficient representing the current for each subsection. Equation (2.54) is then inserted into (2.52).

Equation (2.52) is then multiplied by another series of functions called testing functions. These are also usually simple functions with limited domains. The number of test functions is N, just as for the expansion functions. Both sides of (2.52) are then integrated over the patch surface. Since the testing and expansion functions are nonzero only over a small portion of the patch, this second integration converts the integral equation into a linear system of equations with the α_i as the unknowns as follows

$$[Z][I] = [V] \tag{2.55}$$

where $[Z]$ is an N by N matrix and both $[I]$ and $[V]$ are 1 by N column vectors. $[I]$ contains the α_i coefficients, while $[V]$ contains terms related to the excitation. Explicit expressions for the elements of $[Z]$ and $[V]$ are given in [22]. Equation (2.55) is solved by matrix inversion or any other appropriate technique. The patch current, as given by (2.54), is now known. Any antenna parameter such as radiation pattern and impedance can thus be calculated from the current.

Another rigorous technique that has come into use is the finite element method [23–25], which will only briefly be covered here. Originally developed to solve structural problems, it has been extensively applied to electric machinery, magnetic circuits, scattering, waveguide, and other electromagnetic problems. To solve the microstrip problem, an appropriate differential equation, such as the wave equation, is derived starting from Maxwell's equations. A variational expression (an integral of a quantity called a functional, which is a function of functions) is then constructed from the differential equation and is used to find the function that causes the functional to become stationary, that is achieve either a minima or a maxima. The functional is expressed in terms of a field quantity such as one of the electric field components.

The structure to be analyzed is subdivided into small sections, called elements, which are often triangular in shape. The field quantity of interest is approximated by a polynomial in x and y (z also for three-dimensional problems) within each section. The coefficients of the polynomial are expressed in terms of the field values at the vertices of the element. The vertex field values are unknown and must be determined. There is one polynomial for each section. The polynomials are inserted into the variational expression. The partial derivative of the variational equation is taken with respect to each of the unknown field values and each derivative is set equal to zero. This is the condition that is needed for the functional to be stationary. The result is a linear system of equations in terms of the unknown field values. After solving the system, the field is known everywhere via the polynomial approximations.

A fairly recent addition to the rigorous solution approaches has been the finite difference, time-domain technique [26–28]. This technique directly solves Maxwell's time-dependent curl equations. The derivatives in the curl equations are approximated by central difference formulas. The structure under analysis is covered with a lattice of cubic cells. Points on the cell are assigned to represent one of the components of either the electric or magnetic field. The solution starts by assuming starting values for the field components. Most fields are initially set equal to zero except in the region of the source. The difference equations are time and space stepped through the lattice with previously determined fields at one point in space and at a particular instant of time being used to find the fields at adjacent points and times. The process continues until a steady state is reached.

The finite element and finite difference techniques have not been as widely used for microstrip antenna analysis probably because most problems were already solved using the integral equation method. Generally speaking the integral equation method results in fewer equations to solve, naturally handles the open boundaries of radiation problems, and is readily applied to planar structures like microstrip. Unfortunately, it requires the generation of a Green's function for each problem. Also the Green's functions are singular and highly oscillatory, which makes the numerical integrations very difficult. The other approaches involve easier mathematics and can handle nonlinear and inhomogeneous media. It is harder to model the open boundaries of the radiation problem, and they tend to require more computations and computer memory.

While these rigorous methods are more complete in that they accurately model radiation, surface-wave effects, and the fields internal to the patch, they are quite complicated and not readily implemented especially for design work. Generating a general purpose solver is quite difficult. The programs tend to solve one type or class of problem. Also the accuracy is highly influenced by the feed model. Comparisons presented in Chapter 3 will show that the accuracy of the rigorous approaches may not be much better than the simpler transmission line and cavity models provided the latter are used within their range of validity (for example, electrically thin substrates).

References

[1] Gupta, K. C., R. Garg, and I. J. Bahl, *Microstrip Lines and Slotlines,* Norwood, MA: Artech House, 1979.

[2] Hammerstad, E. O., "Equations for Microstrip Circuit Design," *Proc. European Microwave Conf.,* 1975, pp. 268–272.

[3] Bahl, I. J., and R. Garg, "Simple and Accurate Formulas for Microstrip with Finite Strip Thickness," *Proc. IEEE,* Vol. 65, 1977, pp. 1611–1612.

[4] Kirschning, M., and R. H. Jansen, "Accurate Model for Effective Dielectric Constant of Microstrip with Validity up to Millimeter-Wave Frequencies," *Electron. Lett.,* 1982, Vol. 18, No. 6, pp. 272–273.

[5] Edwards, T. C., *Foundations for Microstrip Circuit Design,* New York, NY: John Wiley & Sons, 1981, pp. 86–88.

[6] Chang, K., ed., *Handbook of Microwave and Optical Components,* Vol. 1, New York, NY: John Wiley & Sons, 1989, pp. 96–107.

[7] Munson, R. E., ''Conformal Microstrip Antennas and Microstrip Phased Arrays,'' *IEEE Trans. on Antennas and Propagation,* Vol. 22, No. 1, Jan. 1974, pp. 74–78.

[8] Derneryd, A. G., ''Linearly Polarized Microstrip Antennas,'' *IEEE Trans. on Antennas and Propagation,* Vol. 24, No. 6, Nov. 1976, pp. 846–851.

[9] Bahl, I. J., ''Build Microstrip Antennas with Paper-Thin Dimensions,'' *Microwaves,* Oct. 1979, pp. 50–63.

[10] Pues, H., and A. Van de Capelle, ''Accurate Transmission Line Model for the Rectangular Microstrip Antenna,'' *IEE Proc. Pt. H (Microwaves, Optics, and Acoustics),* Vol. 131, 1984, pp. 334–340.

[11] Kirschning, M., R. H. Jansen, and N. H. L. Koster, ''Accurate Model for Open End Effect of Microstrip Lines,'' *Elec. Letts.,* Vol. 17, No. 3, Feb. 1981, pp. 123–125.

[12] Derneryd, A. G., ''A Theoretical Investigation of the Rectangular Microstrip Antenna Element,'' *IEEE Trans. on Antennas and Propagation,* Vol. 26, No. 4, July 1978, pp. 532–535.

[13] Jackson, D., and N. Alexopoulos, ''Simple Formulas for the Input Impedance, Bandwidth, and Radiation Efficiency of a Rectangular Patch,'' *IEEE Int. Symp. Digest Antennas and Propagation,* San Jose, CA, 1989, pp. 1130–1133.

[14] Collin, R. E., *Field Theory of Guided Waves,* New York, NY: IEEE Press, 1991, pp. 181–199.

[15] Lo, Y. T., and S. W. Lee, *Antenna Handbook Theory Applications, and Design,* New York, NY: Van Nostrand Reinhold, 1988, pp. 10-10–10-13.

[16] James, J. R., and P. S. Hall, *Handbook of Microstrip Antennas,* Vol. 1, London, UK: Peter Peregrinus Ltd., 1989, pp. 112–120.

[17] Collin, R. E., *Antennas and Radiowave Propagation,* New York, NY: McGraw-Hill, 1985, pp. 273–283.

[18] Carver, K. R., and J. W. Mink, ''Microstrip Antenna Technology,'' *IEEE Trans. on Antennas and Propagation,* Vol. 29, No. 1, Jan. 1981, pp. 2–24.

[19] Rana, I. E., and N. G. Alexopoulos, ''Current Distribution and Input Impedance of Printed Dipoles,'' *IEEE Trans. on Antennas and Propagation,* Vol. 29, No. 1, Jan. 1981, pp. 99–105.

[20] Pozar, D. M., ''Input Impedance and Mutual Coupling of Rectangular Microstrip Antennas,'' *IEEE Trans. on Antennas and Propagation,* Vol. 30, No. 6, Nov. 1982, pp. 1191–1196.

[21] Mosig, J. R., and F. E. Gardiol, ''General Integral Equation Formulation for Microstrip Antennas and Scatterers,'' *IEE Proc. Pt. H (Microwaves, Optics and Acoustics),* Vol. 132, 1985, pp. 424–432.

[22] Chang, K. ed., *Handbook of Microwave and Optical Components,* Vol. 1, New York, NY: John Wiley & Sons, 1989, pp. 783–796.

[23] Sadiku, M. N. O., *Numerical Techniques in Electromagnetics,* Boca Raton, FL: CRC Press, 1992, Chap. 6.

[24] Itoh, T., *Numerical Techniques for Microwave and Millimeter-Wave Passive Structures,* New York, NY: John Wiley & Sons, 1989, Chap. 2.

[25] Silvester, P. P., and R. L. Ferrari, *Finite Elements for Electrical Engineers,* New York, NY: Cambridge University Press, 1983.

[26] Yee, K. S., ''Numerical Solutions of Initial Boundary-Value Problems Involving Maxwell's Equations in Isotropic Media,'' *IREE Trans. on Antennas and Propagation,* Vol. 14, No.3, May 1966, pp. 302–307.

[27] Sadiku, M. N. O., *Numerical Techniques in Electromagnetics,* Boca Raton, FL: CRC Press, 1992, pp. 179–204.

[28] Shibata, T., T. Hayashi, and T. Kimura, ''Analysis of Microstrip Circuits Using Three-Dimensional Full-Wave Electromagnetic Field Analysis in the Time Domain,'' *IEEE Trans.on Microwave Theory and Techniques,* Vol. 36. No. 6, June 1988, pp. 1064–1070.

Chapter 3
Single-Element Design

The cavity model considers a microstrip patch to be a lossy resonator. Virtually any shape that can be considered a resonator has been used or proposed as a microstrip antenna. These include rectangles, circular disks, triangles, annular rings, pentagons, and elliptical disks. For the most part there is little difference in the electrical performance of these various shapes. Some (pentagon and elliptical disk) can be made circularly polarized with only one feed. Others like the annular ring can be excited in higher order modes to radiate patterns with an on-axis null rather than a peak. By far the most popular shape is the rectangle followed by the circular disk. This chapter discusses the design of these two types of radiators. A set of computer programs with examples of designs is presented.

3.1 LISTING OF COMPUTER PROGRAMS

This chapter discusses the basic design of some single-element microstrip patches. Computer programs to assist in analysis and design are also presented. The following is a synopsis of the programs to be covered.

PATCHD: Analyzes rectangular and circular patches. Determines patch size, input resistance at resonance, efficiency, and bandwidth. Useful for trade-off studies. Uses closed-form equations curve fit to numerical full-wave solutions.

PATCH9: Designs rectangular patches. Computes patch length and impedance versus frequency. Incorporates effect of microstrip line or probe feed. Uses transmission line model and closed-form model for feed probe to calculate results.

HWPATCH: Computes rectangular patch radiation patterns in both *E*- and *H*-planes. Includes the effect of dielectric substrate. Uses an assumed patch surface current distribution in an approach based upon reciprocity.

SCPATCH: Designs quarterwave short-circuited patches. Computes patch length and impedance versus frequency. Incorporates effect of feed. Also models the use of shorting pins. Essentially a modified version of PATCH9.

PATCHC: Designs rectangular patches with a contacting cover layer. Computes patch length and impedance versus frequency. Incorporates feed effects. A modified transmission line model. Uses a variational technique to include cover layer.

CHWPATCH: Computes *E*- and *H*-plane radiation patterns for a covered patch including substrate and cover layer effects. Uses a modified version of the model in HWPATCH.

CPATCH: Designs circular patches. Computes patch diameter and impedance versus frequency. Includes effects of a probe feed, if used. Uses closed-form equations curve fit to numerical full-wave solutions. Models patch as a parallel RLC circuit.

CIRPAT: Computes circular patch *E*- and *H*-plane radiation patterns. Uses asymptotic equations derived from full-wave solutions.

WRAPPAT: Designs wraparound patches. Determines patch length, number of feeds, and impedance at each feed. Uses a transmission line model.

WRAPRND: Calculates radiation pattern for a wraparound patch including effect of substrate. Uses a Fourier transform domain solution to the wave equation with an assumed surface current distribution to find pattern.

3.2 SUBSTRATE SELECTION

One of the first tasks in patch design is the selection of a suitable substrate material. The major electrical properties to consider are relative dielectric constant and loss tangent. A high dielectric constant results in a smaller patch but generally reduces bandwidth and results in tighter fabrication tolerances [1]. A high loss tangent reduces antenna efficiency and increases feed losses. Generally it is best to select a substrate with the lowest possible dielectric constant consistent with the space available for the antenna [2]. Substrate thickness should be chosen as large as possible to maximize bandwidth and efficiency, but not so large as to risk surface-wave excitation. For a maximum operating frequency of f_u, the thickness should satisfy [3]

$$h \leq \frac{0.3c}{2\pi f_u \sqrt{\epsilon_r}} \tag{3.1}$$

where c is the speed of light, 3×10^8 m/s, and ϵ_r is the relative dielectric constant. If the system must operate within a narrow frequency band, dielectric constant stability is important. A variable ϵ_r will cause the patch frequency to wander, which can reduce product yield.

The intended operating environment plays an important role in substrate selection. Some substrates warp or creep under high temperatures. Dimensional changes with temperature alter electrical performance and can degrade reliability. Other mechanical considerations include machinability and flexibility. Substrate mechanical properties such as operating temperature range, thermal coefficient of expansion, thermal conductivity, and flexibility or rigidity should be checked against system requirements. Finally and often most important, trade-offs between cost and performance must be made.

Traditional patch design at microwave frequencies uses substrates such as alumina, quartz, and PTFE. These offer excellent electrical performance, but resulting substrate costs are often too high for commercial applications. Epoxy/glass FR4 is a widely used material for low-frequency and digital circuit boards. Its low cost and ease of processing are its strengths. A high loss tangent and relatively variable dielectric constant limit usage to frequencies below about 1 GHz or so. Low-cost commercial RF/microwave designs need substrates that combine the advantages of both of the above.

Recently substrate manufacturers have introduced new materials that provide good electrical performance at reasonable prices. Table 3.1 lists some of these new substrates but is certainly not all inclusive. The loss tangents are comparable to standard microwave materials. The cost for the new materials is three or more times less than traditional microwave substrates and, in some cases, approaches those of epoxy/glass. For comparison, FR4 is also listed. Note its high loss tangent.

Table 3.1
New Low-Cost Substrates for Microstrip Patch Antennas

Material	*Relative Dielectric Constant*	*Loss Tangent*	*Manufacturer*
RO3003	3.00	0.0013	Rogers Corp.
RO3006	6.15	0.0013	Rogers Corp.
RO3010	10.2	0.0013	Rogers Corp.
RO4003	3.38	0.0022	Rogers Corp.
TLC-32	3.2	0.003	Taconic Plastics
HT-2	4.3	0.0033	Hewlett-Packard
Polyguide	2.32	0.0005	Shawinigan Research
FR4	4.4	0.01	

For low-frequency RF applications, those having modest RF performance requirements, or systems with a very limited RF section, FR4 may be the best choice. With FR4, the RF section can be easily integrated with other electronics. No new processing has to be introduced and cost will be minimal. For most systems operating above 1 GHz or so and those with significant electrical performance specifications, the substrates of Table 3.1 should be considered. At very high microwave (10 GHz or so) and millimeter-wave frequencies (> 30 GHz), the PTFE, alumina, or quartz materials may have to be used.

3.3 RECTANGULAR ELEMENT ANALYSIS AND TRADE-OFFS

To design a rectangular patch, one must (i) choose a substrate, (ii) select a feed approach, (iii) determine the correct length, (iv) decide on a patch width, and (v) find the feed location. Substrate selection has already been discussed. For the other aspects, there are several options and trade-offs. These are covered in this section.

Rectangular patches can be fed in a variety of ways. Figure 3.1 shows two of the most common. For both types, the feed is placed in the center of the patch with respect to its width, *W*. The patch is fed along the direction of resonance, that is parallel to its length, *L*. Other feeds are possible including several where there is no contact between the feed and the patch. Two of these are discussed in Chapter 4.

In Figure 3.1(a), a microstrip line feeds the patch. The line can either end at the patch edge or be inset as shown. The amount of inset, *l* in the figure, is chosen to match the patch impedance with that of the feed line. If the line terminates at the patch edge, a quarterwave transformer is used for matching. Microstrip line feeding is perhaps the

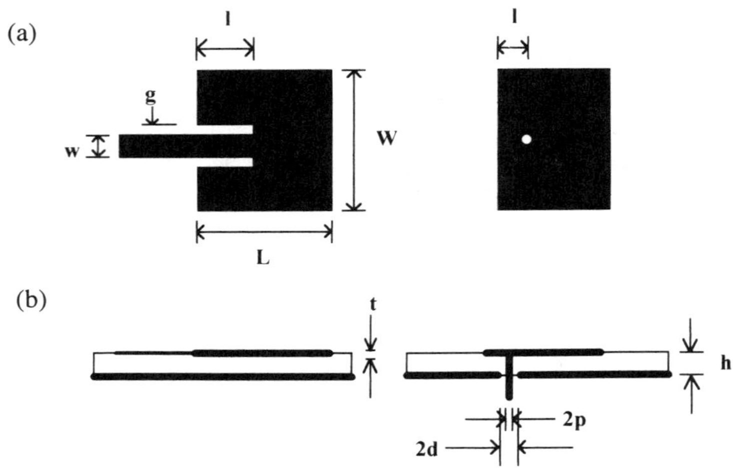

Figure 3.1 Two common microstrip patch feeds: (a) microstrip feed and (b) probe feed.

Rectangular Patch Design

INPUT ANTENNA FREQUENCY (GHz)? **1.575**

When appropriate, input prompts indicate the required units. In this case, the frequency must be given in gigahertz.

The next input prompt is

INPUT SUBSTRATE RELATIVE DIELECTRIC CONSTANT AND LOSS TANGENT? **2.2, 0.001**

We have chosen parameters that are typical for a PTFE-type substrate. The curve-fit formulas are valid for $1 < \epsilon_r < 10$. If an ϵ_r value outside this range is input, the program will print a warning message and then ask if you want to continue or re-enter another value. The program now asks if you want to analyze a square or rectangular patch.

DO SQUARE (s) OR RECTANGULAR PATCH (r)? **r**

If you answer with **r**, the next input is

INPUT PATCH WIDTH (cm)? **9.434**

A 9.434-cm width will turn out to be about 1.5 times the patch length. The program does an initial estimate of patch length and checks the width-to-length ratio to see if it is between 0.9 and 2.0, the model validity limits. If not, it again asks if you want to change inputs or continue. The substrate height, h, in Figure 3.1, is then requested by

INPUT SUBSTRATE HEIGHT (cm)? **.16**

Two checks against the model range are now done. One is the height-to-length ratio, which should be less than 0.2. The other is $\sqrt{(\epsilon_r - 1)}h/\lambda_0 \leq 4$, which must be satisfied. ϵ_r is the substrate relative dielectric constant, and λ_0 is the free-space wavelength. The option to change or not is repeated. Next the program requests the metalization conductivity relative to copper by

INPUT CONDUCTOR CONDUCTIVITY RELATIVE TO COPPER? **1**

The impedance versus frequency behavior limits the patch bandwidth. Bandwidth is a function of the patch quality factor (Q) and the degree of mismatch that can be tolerated. For this case, assume that a standing wave ratio (SWR) of 2.0:1 is acceptable over the operating frequency range.

INPUT ACCEPTABLE swr FOR BANDWIDTH CALCULATION? **2**

The final input is the feed location, which can be from the patch edge ($l = 0$) to half way in ($l = L/2$). The impedance goes from its highest at the edge to zero at the patch center following a cosine-squared variation with inset.

INPUT FEED LOCATION?
[0 < lf < 3.211]
(patch edge patch center)
0.0

3.211 is half the estimated patch length. We are not insetting the feed in this example.

PATCHD now computes the patch length. The program does this by evaluating the curve-fit formulas, which calculate the complex resonant frequency given the patch dimensions and substrate properties. The length is iterated until the real part of the calculated complex frequency agrees with the input frequency within 0.001 GHz. PATCHD performs up to 100 iterations. If the resonant length is not found, the 0.001 GHz tolerance limit can be relaxed by the user.

Efficiency and bandwidth are determined, in part, from the patch Q. The Q has several components. Two contributors are the losses in the conductor and dielectric (Q_c and Q_d), and expressions for these are given in Chapter 2. The power radiated by the space and surface waves also contribute. Their combined Q is found from the complex resonant frequency, $\omega_c = \omega_r + j\omega_i$, by

$$Q_r = \frac{\omega_r}{2\omega_i} \tag{3.2}$$

Separation into individual radiation and surface Q factors can be done using the radiation efficiency calculation given in [6]. In [6] Pozar finds closed-form expressions for the space- and surface-wave powers radiated by an elemental dipole on the surface of a grounded dielectric slab. The space-wave power is

$$P_{sp} = \frac{\eta_0 k_0^2 k_{0d}^2}{3\pi}\left[1 - \frac{1}{\epsilon_r} + \frac{2}{5\epsilon_r^2}\right] \tag{3.3}$$

where $k_0 = 2\pi/\lambda_0$, $k_{0d} = k_0 h$, and $\eta_0 = 377\Omega$. The surface-wave power is

$$P_{sur} = \frac{\eta_0 k_0^2}{4} \frac{\epsilon_r(x_0^2 - 1)}{\epsilon_r[1/\sqrt{x_0^2 - 1} + \sqrt{x_0^2 - 1}/(\epsilon_r - x_0^2)] + k_{0d}[1 + \epsilon_r^2(x_0^2 - 1)/(\epsilon_r - x_0^2)]} \tag{3.4}$$

where

$$x_0 = 1 + \frac{x_{0n}}{x_{0d}},\ x_{0n} = -\epsilon_r^2 + \alpha_0\alpha_1 + \epsilon_r\sqrt{\epsilon_r^2 - 2\alpha_0\alpha_1 + \alpha_0^2},\ \text{and } x_{0d} = \epsilon_r^2 - \alpha_1^2$$

with $\alpha_0 = s\tan(k_{0ds})$, $\alpha_1 = -[\tan(k_{0ds}) + k_{0ds}/\cos^2(k_{0ds})]/s$, $k_{0ds} = k_{0d}s$, and $s = \sqrt{\epsilon_r - 1}$. It is shown in [7] that (3.3) and (3.4) closely approximate the values for a patch irrespective of its shape.

The sum of P_{sp} and P_{sur} gives the total radiated power. The ratio of space-wave power (which is the desired radiation) to the total power determines the radiation efficiency

$$\eta = \frac{P_{sp}}{P_{sp} + P_{sur}} \tag{3.5}$$

From the efficiency and Q_r, the space- and surface-wave Q factors are determined as

$$Q_{sp} = \frac{Q_r}{\eta} \tag{3.6}$$

$$Q_{sur} = \frac{Q_r}{1 - \eta} \tag{3.7}$$

The total patch quality factor, Q_{total}, is the sum of the inverses of the above Q factors, denoted

$$\frac{1}{Q_{total}} = \frac{1}{Q_{sp}} + \frac{1}{Q_{sur}} + \frac{1}{Q_c} + \frac{1}{Q_d} \tag{3.8}$$

The overall efficiency is the radiated power divided by the sum of the space-wave, surface-wave, conductor-loss, and dielectric-loss powers. Since the stored energy is the same for all Q factors, these powers are proportional to the Q's. Then

$$\eta_{tot} = \frac{Q_{total}}{Q_{sp}} \tag{3.9}$$

The acceptable SWR and Q_{total} [8] determine the bandwidth, BW. It is given by

$$BW = \frac{\text{SWR} - 1}{Q_{total}\sqrt{\text{SWR}}} \tag{3.10}$$

which shows the trade-off between bandwidth and SWR.

The next step is to determine the patch input resistance. Near its resonant frequency, the patch impedance can be approximated by a parallel RLC circuit. Abboud et al. [9] show with this model that the input resistance at resonance is

$$R = \frac{Q_{total}}{2\pi f_r C_p} \tag{3.11}$$

with f_r being the resonant frequency and

$$C_p = \frac{\epsilon_r \epsilon_0 LW}{2h \cos^2\left(\frac{\pi l_f}{L}\right)} \tag{3.12}$$

with $\epsilon_0 = 8.854 \times 10^{-12}$ F/m. Note that C_p is the capacitance of a rectangle having the same dimensions as the patch but with an additional factor due to feed inset.

After performing the calculations, PATCHD again clears the screen and displays the results

```
                PATCHD.V10   11-04-1995     17:27:59

SUBSTRATE HEIGHT = 0.160 cm
SUBSTRATE RELATIVE DIELECTRIC CONSTANT = 2.20
SUBSTRATE LOSS TANGENT = 0.0010
CONDUCTOR RELATIVE CONDUCTIVITY = 1.000
PATCH LENGTH = 6.422 cm
PATCH WIDTH = 9.434 cm
FEED LOCATION = 0.000 cm
FREQUENCY = 1.5749 GHz

      INPUT RESISTANCE = 151.73 Ohms
      PATCH TOTAL Q = 53.712
      EFFICIENCY = 96.89%
      OVERALL EFFICIENCY = 86.27%
      PATCH BANDWIDTH = 1.32%
      FOR A 2.00:1 SWR
```

These results are typical for patch antennas. Edge feed antennas exhibit fairly high resonant resistances. Surface-wave excitation, as seen by the 96.89% efficiency, is small if the substrate height is less than about 0.01λ. Element overall efficiency is good. Feed line losses usually account for the poorer efficiency of patch arrays. Bandwidths are on the order of 1% to 3% for thin substrates.

To get an idea of the trade-offs involved with the various design parameters, the 1.575-GHz patch example was re-run using different values. The results are shown in Table 3.2. The patch width affects input resistance, total efficiency, and bandwidth. A wider patch has a lower input resistance, higher efficiency, and a wider bandwidth. Thin substrates have lower efficiencies, smaller bandwidths, and lower input impedances. Thicker substrates produce wider bandwidths but have higher surface-wave levels and consequently lower radiation efficiencies. As the dielectric constant rises, so does the resistance and surface-wave level, but the efficiency and bandwidth are reduced. The patch length depends mostly on dielectric constant ranging from 8.788 cm for $\epsilon_r = 1.1$ to 4.470 cm when $\epsilon_r = 4.4$. Substrate height has a secondary effect. In this example, the length reduces by about 9% as the height goes from 0.0635 cm to 0.635 cm.

Table 3.2
Effect of Various Design Parameters on Patch Characteristics

Substrate ϵ_r	*Substrate Height (cm)*	*Patch Width (cm)*	*Input Resis. (Ohms)*	*Radiation Efficiency (%)*	*Total Efficiency (%)*	*Bandwidth (%)*
2.2	0.16	6.096	300.3	96.9	83.2	1.02
2.2	0.16	9.398	151.7	96.9	86.3	1.32
2.2	0.16	13.462	92.8	96.9	87.6	1.51
2.2	0.0635	9.398	109.4	98.7	63.2	0.71
2.2	0.635	9.398	167.9	88.7	87.2	5.15
1.1	0.16	13.208	146.4	99.9	89.4	1.37
4.4	0.16	6.604	212.0	93.2	78.7	0.93

3.4 RECTANGULAR ELEMENT DESIGN

Although PATCHD determines patch length and input resistance for a given set of inputs, a more detailed design procedure is often appropriate. This can be done using program PATCH9. PATCH9 uses a transmission line model [3] to determine patch length and the input impedance versus frequency including the effects of the feed. A description of this model is given in Section 2.3.

The program has both a design and analysis mode. First we will look at its operation in the design mode. To see how it is used, let us apply it to the design of the 1.575-GHz patch of the previous section. The following shows what is displayed as the program executes along with the inputs for the design example. Upon invoking PATCH9, the screen appears as follows.

PATCH9.NEW 3/15/95 2:45:00
[Current Program Date - 2/21/95]

Rectangular Microstrip Patch Design & Analysis

Do design (d) or analysis (a)? **d**

Input design frequency in GHz? **1.575**

Input relative dielectric constant? **2.2**

Input loss tangent? **.001**

Input substrate height (cm)? **.16**

(Transmission line models are only valid for electrically thin substrates. This limit is usually taken to be 0.01λ. The program warns if this height is exceeded and allows for re-entering of substrate height. No change is required, but one should be aware of potential model accuracy degradation.)

simplest and costs the least. It makes for a completely printed structure. A disadvantage is potential feed radiation.

The patch can also be fed with a probe through the ground plane as in Figure 3.1(b). The probe could be the center conductor of a coaxial cable or connected to a microstrip line on a substrate behind the ground plane. The probe feed is easily inset for matching. Insetting also minimizes probe radiation. The ease of insetting and reduced radiation are advantages of probe feeding. Another advantage results from the increase in available space for feeds located on another substrate.

The patch length is determined by the condition for resonance. This occurs when the input impedance is purely real. L is about 0.48 to 0.49 of the wavelength in the dielectric. The patch width generally lies between about 0.5 to 2.0 times the length. The width can be used to vary the input impedance with wide patches having lower impedances but should not exceed one wavelength to avoid higher order transverse modes. For narrow patches, the radiation efficiency becomes low. Also the microstrip feed can perturb narrow patch operation by "blocking" the radiating edge.

Often it is necessary to do "what-if" or trade-off studies before proceeding with an actual design. The designer may want to look at several possible substrates or investigate the effect of a patch dimension. These studies can be performed using program PATCHD. PATCHD analyzes rectangular and circular patches. It uses curve-fitted formulas derived from full-wave solutions [4,5]. As such, the results include the effects of radiation and surface waves. They do not, however, include any feed interactions other than feed inset. The program determines the patch length or diameter, input resistance at resonance, efficiency, and bandwidth.

As an example, PATCHD is applied to study options for a rectangular patch at 1.575 GHz, which is one of the frequencies of the *global positioning system* (GPS). GPS is a constellation of satellites that provide position information virtually anywhere in the world. GPS is used for navigation and location. PATCHD is an executable program and so can be run in the usual manner by typing PATCHD at the DOS prompt or "clicking" on it from the File Manager in Windows. When invoked, the program first clears the screen and then prints some header lines followed by a request for the patch shape to analyze. The screen should look as follows (italics indicate what the program displays on the screen and bold type shows what the user types).

PROGRAM PATCHD 5/13/95 2:45:00
[Current Program Date - 5/9/95]

DO RECTANGULAR (r) OR CIRCULAR (c) PATCH? **r**

Since we are interested in a rectangular patch, the appropriate response is **r**. For most of the programs, the inputs are case insensitive so either **r** or **R** will work. Also for most programs, error traps are built in so if a wrong letter is typed, the input request will be repeated. After typing in **r**, the following is displayed:

Input line thickness (cm)? **.00356**

Input line conductivity relative to copper? **1**

Input patch length (cm)

Estimated patch length = 6.421 cm ?

6.421

[PATCH9 estimates patch length by calculating $\lambda_0/(2\sqrt{\epsilon_r})$.]

Input patch width (cm)? **9.434**

(Program checks to see if patch width is $\leq \lambda$ in dielectric and provides an option to change width.)

Input feed type: m = microstrip, p = probe, n = none? **m**

(We will discuss feeds after this example. For now we assume a microstrip line feed with no inset.)

Input feed point distance wrt to patch edge (cm)? **0.0**

Input feed line width (cm)? **.495**

(Width for a 50-Ω line. To warn of potential feed blockage, the program signals if the feed width exceeds 0.25 the patch width. There are no guidelines for this so the limit is somewhat arbitrary.)

Input fractional change (%) in patch length for initial search
(default = 5%)

? <**rtn**>

PATCH9 finds the resonant length by iterating it until the calculated input reactance is zero. The iteration is performed by calculating the impedance and then either decreasing or increasing patch length depending upon the sign of the reactance. The default length increment is 5%. By hitting the enter key, the default is chosen. In some cases this is too large a change and the model will not converge on a solution. This is particularly true for thick substrates or if the input resonant length is far off the true value. For these situations, input a smaller percentage value. If the program cannot find a solution after 100 iterations, it stops and asks if the user wants to try another patch length or quit looking for the length.

Once the length is found, PATCH9 is ready to compute the input impedance as a function of frequency.

Input start and stop freqs (GHz) for impedance calc? **1.55,1.60**

(Since the patch bandwidth is small and the transmission line model is only valid near the lowest order mode resonance, the calculation frequency bandwidth should be kept to within a few percent.)

Input number of frequencies? **10**

Using the transmission line model and the calculated resonant length, the program finds the input impedance at each of the requested frequencies.

When completed calculating, the screen is cleared and the output results displayed.

PATCH9.NEW 11-04-1995 17:46:00
Current Program Date - 05/06/1995

Substrate height (cm) = 0.160
Relative dielectric constant = 2.200
Loss tangent = 0.0010
Line thickness (cm) = 0.00356
Patch width (cm) = 9.434
Patch length (cm) = 6.293
Feed type = microstrip line
Feed point inset (cm) = 0.0000
Feed line width (cm) = 0.495
Resonant frequency (GHz) = 1.575000
Resonant impedance = 160.665 j 0.0039 ohms

FREQUENCY (GHz)	INPUT IMPEDANCE (ohms)
1.550	33.099 J 65.223
1.555	46.443 J 73.086
1.560	67.532 J 79.568
1.565	99.757 J 78.219
1.570	139.515 J 54.601
1.575	160.665 J 0.004
1.580	139.265 J –54.296
1.585	99.755 J –77.644
1.590	67.826 J –79.038
1.595	46.888 J –72.716
1.600	33.598 J –65.007

Notice that the resonant length and resonant input resistance values are different from those determined by PATCHD. This is because (i) PATCHD does not include the presence of the feed (except for the inset) and (ii) the models are different. The impedance can also be plotted on a Smith Chart and displayed on the screen in response to the prompt

PLOT RESULTS ON SMITH CHART? **y**

Figure 3.2 shows the impedance when plotted on a Smith chart. At this point, the program is waiting for any input to continue. Just strike the enter key to continue.

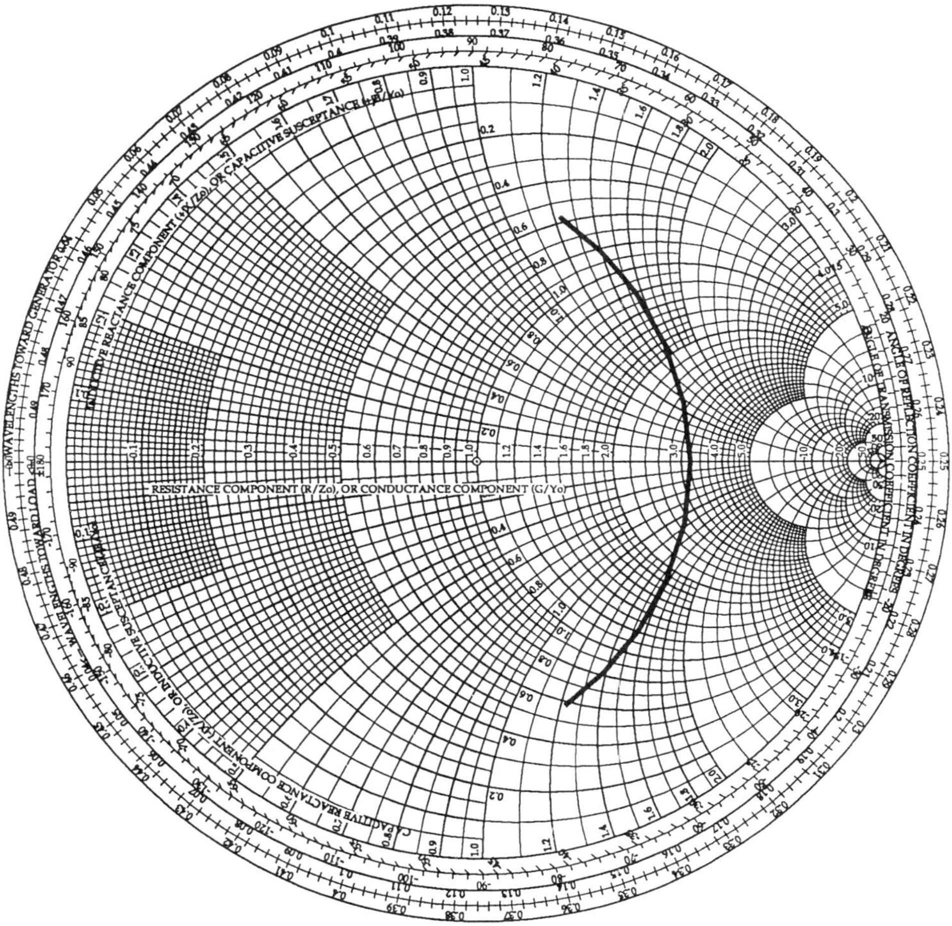

Figure 3.2 Program display of computed patch impedance on a Smith chart.

The output can now be printed out if desired for the next prompt from the program is

PRINT OUT RESULTS (y or n)? **n**

Finally the user is given the option of storing the output in a file via

STORE RESULTS IN A FILE (y or n)? **y**

When answered with a yes, the data can be stored as displayed on the screen or in a TOUCHSTONE (EESOF TM) compatible file.

STORE DATA IN A TOUCHSTONE COMPATIBLE FILE (y or n)? **n**

INPUT NAME OF FILE TO STORE RESULTS? **patch9.out**

At this point, the data is stored as it was displayed on the screen in the file patch9.out. If storage in a TOUCHSTONE compatible file is requested, the inputs become

NAME OF FILE (no extension) TO STORE RESULTS? **patch9**

(PATCH9 automatically puts the extension .S1P on the file name.)

The user has the opportunity to put two header lines in the file via

INPUT (optional) HEADER LINE FOR FILE? **1.575 GHz Patch Design**

INPUT (optional) ADDITIONAL LINE FOR FILE? **GPS Locator Project**

In addition to the header lines, PATCH9 writes the identifier line that tells TOUCHSTONE about the format of the data. This line gives the frequency units and the facts that the data are *Z*-parameters and the impedance is in rectangular form and normalized to 1Ω. This line looks like # *GHZ Z RI R 1*. To make the file more readable, one final line is included, that is a header over the data to label the columns as ! *FGHZ REAL[Z] IMAG[Z]*.

The last input is

DO ANOTHER CASE (y or n)? **n**

which allows the user the chance to either exit or do another calculation.

PATCH9 has three options for feed. One is a microstrip line as in the above example. When this option is chosen, the feed line effective width is calculated as indicated in Chapter 2 [10]. It is then subtracted from the patch effective width on the edge being fed. The edge slot conductance and susceptance are therefore reduced for that edge. This accounts for the ''shadowing'' of the radiating edge by the feed.

When insetting the microstrip line, some gap, as g in Figure 3.1(a), must be used to isolate the line from the patch. There are no particular guidelines for this. As an approximate rule-of-thumb, a spacing of 0.10 to 0.50 of the line width is suggested. Wide lines have more fields concentrated under the line; thus narrower spacings are appropriate. Conversely narrow lines should have wider spacings. For all cases, the combination of feed and gap width should be less than about 0.25 the patch width to avoid blockage.

The introduction of the gap changes both the resonant frequency and resonant resistance. Unfortunately there is little published data on the gap influence. The effect on resonant frequency is small. It increases (from the value with no gap) almost linearly with gap width for widths up to the feed line width. The maximum frequency change is slightly less than 1%. The resistance is effected much more dramatically. The resistance decreases in an approximately linear fashion with increasing gap width. When the gap width approaches the feed line width, the resistance has decreased by about 60%. Since both changes appear to be linear, it is possible to estimate their effect when designing patches with inset feeds. In particular, the change in resistance should be accounted for.

Another feed type that can be analyzed is the probe feed. To use a probe feed answer **p** when PATCH9 asks for feed type. Input the feed inset distance at the next prompt. The probe should be inset from the patch edge. If not, it may radiate and corrupt the patch pattern. The minimum inset would place the coaxial relief hole (2d in Figure 3.1(b)) just beneath the patch.

For a probe feed, PATCH9 requests the probe dimensions,

Input feed probe diameter (cm)? **.0635**

(see 2p in Figure 3.1(b)).

Input feed probe relief hole diameter (cm)? **.1460**

The probe model [11] in the program has validity limits. For some like dielectric constant ($1 < \epsilon_r < 10$) and substrate thickness [$(kh)^2 \ll \pi$, $k = 2\pi(\sqrt{\epsilon_r})/\lambda_0$], only a warning is printed. To change these inputs, the program must be rerun. If the relief hole diameter is too large [$(kd/2)^4 \ll 1$], both the probe and relief hole diameters can be re-entered. The program then computes the equivalent circuit elements for the probe. These are included in the patch impedance evaluation.

The probe model is based upon an analysis of a coaxial line with its outer conductor flush mounted to the lower conducting plane of a parallel plate waveguide. The inner conductor extends up to the upper plate. A TEM field distribution is assumed across the coaxial aperture. The various parallel plate modes are found, and the probe admittance is determined by enforcing a continuity of power flow at the aperture. Under the validity limits described earlier, the results can be reduced to closed form and produce the equivalent circuit of Figure 3.3.

The equivalent circuit elements are

$$R_p = \frac{\omega \mu_0 h}{4} \tag{3.13}$$

$$L_p = -\frac{\mu_0 h}{2\pi}\left[\ln\left(\frac{kp}{2}\right) + \gamma\right] \tag{3.14}$$

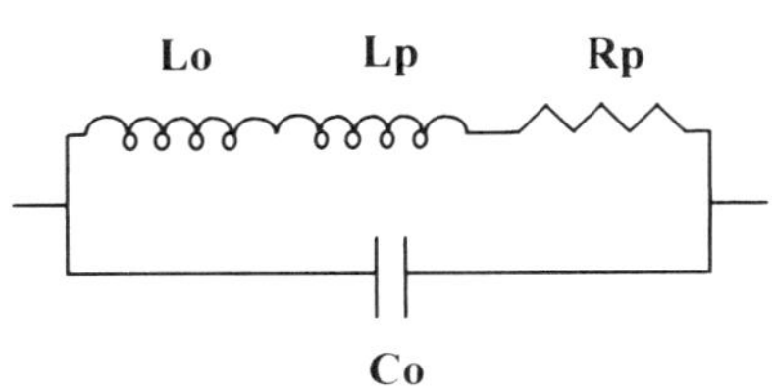

Figure 3.3 Probe feed equivalent circuit.

$$C_0 = \frac{\epsilon_0 \epsilon_r}{6h \ln^2\left(\frac{p}{d}\right)} \left\{ 3\pi \left[d^2 - p^2 - 2d^2 \ln\left(\frac{p}{d}\right) \right] \right.$$

$$\left. + 4\pi h^2 \ln\left(\frac{p}{d}\right) - \frac{12h^3}{\pi^2 d} \cdot \left(1.202 - \sum_{n=1}^{\infty} n^{-3} \exp\left[\frac{-2n\pi(d-p)}{h}\right]\right) \right\} \tag{3.15}$$

$$L_0 = \frac{-\mu_0 h k^2}{4\pi \ln\left(\frac{p}{d}\right)} \left[\left(d^2 + p^2 \ln\left(\frac{p}{d}\right) - d^2 + p^2 \right] \left[\ln\left(\frac{k_1 p}{2} + \gamma\right] \right. \right. \tag{3.16}$$

with $\omega = 2\pi f$, $\mu_0 = 4\pi \times 10^{-7}$ H/m, and $\gamma = 0.5772$ (Euler's constant). The predictions of the equivalent circuit are shown in [11] to agree quite well with a rigorous integral equation solution, provided the probe is not located very close to the patch edge.

The final feed option is to have no feed (**n**). The transmission line model is evaluated with no feed effects. Only a feed inset may be included.

3.5 COMPARISON TO MEASURED RESULTS

In the above example it was noted that the calculated patch length and input resistance were different depending upon the model. Transmission line models, while easily implemented, have limited accuracy. It is important to have an idea about the accuracy of PATCH9. Reference [12] presents a series of measurements of patches with both microstrip and probe feeds. Also shown are the predictions of cavity and full-wave models. To assess its validity, PATCH9 was run in its analysis mode (described in the next section) to calculate resonant frequency and resistance for the patches of cases 1 through 8a of [12]. The substrate thickness for case 8a is quite large being 0.04λ.

The resonant frequency predicted by PATCH9 had an average error of 2.18%. This compares with 1.38% for the full-wave solutions. Interestingly, PATCH9 did much better for the probe-fed cases with a typical error of less than half that of microstrip feeds. Except for case 8a, the average error for resonant resistance is 16.2% with PATCH9 compared to 17.5% for the full-wave solution. Both had errors of over 100% for case 8a. The PATCH9 predictions for resistance are better for the microstrip fed cases. From these comparisons, it can be seen that PATCH9 gives reasonable results considering the computational effort.

3.6 PATCH9 ANALYSIS MODE

PATCH9 can also be used to analyze rectangular patches in much the same manner as PATCHD. When the first prompt is answered with an **a**, PATCH9 asks for the same inputs as before. Instead of iterating the patch length, it simply evaluates the transmission line model for the given length. The result is printed out along with a request for the next action. Using the same inputs as the last section, we get

Patch Length (cm) = 6.421
Frequency = 1.5750 GHz
Z = 23.0373 j −56.8893

Input new frequency (f), patch length (l), patch width (w) or analyze (a)?

Z is the impedance for the input values. The user has the option of changing the frequency (that is, find the frequency for a given length), the patch length (find length for the frequency), patch width (change the input impedance magnitude), or continue with the given inputs. If the latter is selected, PATCH9 asks for frequency data, calculates impedance versus frequency, and then displays output data as in the design mode.

3.7 RECTANGULAR PATCH RADIATION PATTERNS

The patch design procedure has concentrated on finding the proper length and adjusting the width and feed positions to obtain the desired input resistance. This is due to the narrow impedance bandwidth. Patch patterns are relatively constant over a much wider bandwidth. They are slow functions of the patch dimensions and substrate properties. Although the pattern variations are not large, some "tweaking" may be required for certain applications.

The factors that effect the patterns are the patch width, substrate dielectric constant, and, to a lesser degree, substrate height. The antenna polarization is linear with the E-field parallel to the patch length. Therefore the E-plane lies along the length, while the H-plane is parallel to the width. Figure 3.4 gives H-plane patterns for a square patch (W/L = 1.00) and a wavelength wide patch (W/L = 2.03). In the patterns, 0^0 is broadside (perpendicular) to the patch. The H-plane pattern beamwidth is a function of patch width. As described in Chapter 2, the intuitive model has the radiation occurring from two slots located at each end of the patch. As the patch is widened, the slots get longer. The longer aperture has a narrower pattern. The E-plane is not effected by patch width.

E-plane patterns for two different dielectric constants are given in Figure 3.5. The E-plane pattern widens with increasing dielectric constant. Again the radiating slot model offers an explanation. The patch length shrinks with increasing dielectric constant. This brings the radiating edges closer together. The effective size of the array formed by the two edges becomes smaller, thereby producing a wider pattern. The E-plane pattern is also slightly affected by substrate height. It shows up as a faster pattern roll-off at very wide pattern angles, > 80, degrees with thicker substrates.

These patterns are calculated assuming an infinite ground plane. Because the patterns are broad, they are affected by "real world" finite-sized ground planes, ridges, and bumps. The energy radiated along the ground plane (~ 90 degrees) can be scattered in many

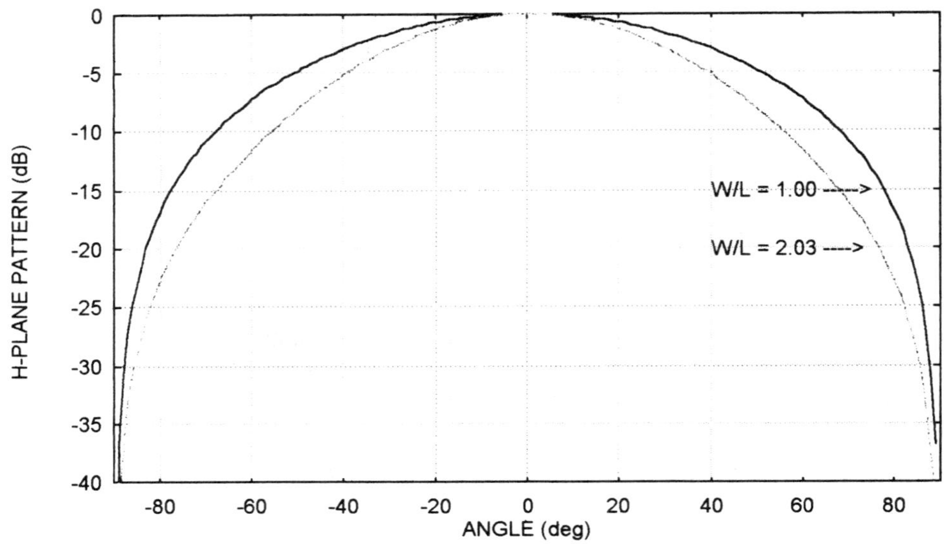

Figure 3.4 Rectangular patch *H*-plane pattern as a function of width.

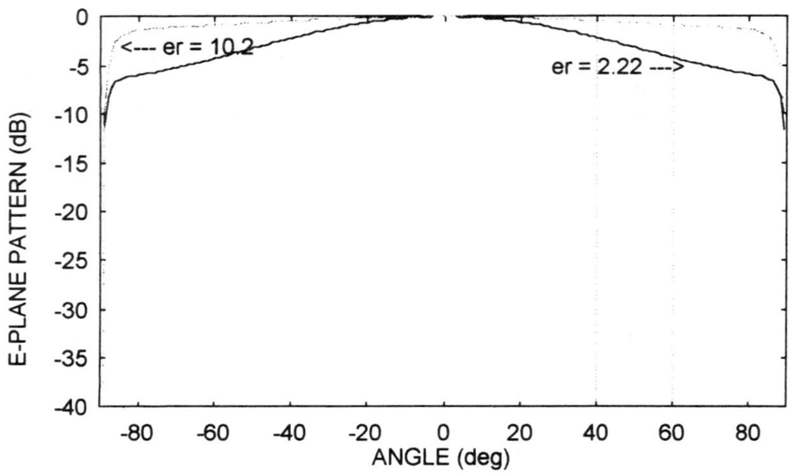

Figure 3.5 Rectangular patch *E*-plane pattern for two dielectric constants.

directions by the edges. It then combines with the direct radiation from the patch. For very small (< λ_0) ground planes, ripples are introduced into the pattern over a wide range of angles. Also much radiation spills over onto the backside of the ground plane. This creates a very broad pattern.

As the size of the ground plane increases, the ripples in the main pattern diminish. With a several-wavelength-wide ground plane, the *E*-plane pattern displays a more gradual roll-off than seen in Figure 3.5. The *H*-plane pattern does not go to zero but gradually tapers to about the –20-dB level. Backlobe radiation behind the ground plane is typically about 15 dB to 20 dB below the pattern peak.

The effect of a finite ground plane or ridges can be quite accurately predicted using the *geometrical theory of diffraction* (GTD). The application of GTD to patches is presented in [13] where good agreement between theory and measurements occurs. A good discussion of GTD and how to apply it is given in [14]. Reference [14] also contains computer programs for the diffraction coefficients employed in [13].

Patch patterns for an infinite ground plane can be found using HWPATCH. This program uses the approach of [7], which is described in Chapter 2. The presence of the substrate is included. The patch surface current density is taken to be cosinusoidal, which means that the length should be close to the resonant value. HWPATCH assumes that the patch length and width are already known. When executed, the screen looks like

HWPATCH.BAS 5/22/95 09:35:00
[Current Program Date - 20/5/95]

Calculates patch antenna far field pattern using electric current model. Does both E- and H-planes. Stores results in user input files.

INPUT PATCH LENGTH (cm)? **6.292**

INPUT PATCH WIDTH (cm)? **9.434**

INPUT PATCH SUBSTRATE ER? **2.22**

INPUT PATCH SUBSTRATE HEIGHT (cm)? **.0790**

INPUT FREQUENCY (GHz)? **1.575**

INPUT START AND STOP ANGLES FOR PATTERN CALCULATION?
(broadside = 0 deg) **–90,89**

(Since the calculation assumes an infinite ground plane, angles < –90 degrees and > +90 degrees are behind the ground plane where no field exists. Any desired angle range –90 degrees < θ < 90 degrees can be input.)

INPUT NUMBER OF ANGLES FOR PATTERN CALCULATION? **179**

(With this input, the pattern will be determined at 1-degree increments. Up to 720 angles can be calculated.)

After computing the pattern, the program asks

WANT TO PLOT PATTERN (y or n)? **y**

If answered yes, the next prompt is

RECTANGULAR (r) OR POLAR (p) PLOT? **r**

For rectangular plots, the angle range of the plot must be specified via

INPUT ANGLE SCALE (45,90,180)? **90**

Only three types of plots are possible with angle ranges of ±45 degrees, ±90 degrees, and ±180 degrees. The built-in plotting routine works best when the previously input start and stop angles agree with these ranges. It will plot O.K. if the input angle range is less than the above scales, but the grid will extend beyond the data. For arbitrary ranges, the pattern can be stored in a file and plotted with an external routine. No input of scale is needed for polar plots. The screen is cleared and the plot displayed.

Figure 3.6 shows representative plots for both types. The solid line is the *E*-plane pattern, while the dots are that of the *H*-plane. To continue, hit the enter key. The user is given the option of storing the patterns in files.

STORE E- AND H-PLANE PATTERNS IN FILES (y or n)? **y**

When answered yes, the screen prompts

INPUT FILENAME (no extension) TO STORE PATTERNS? **hwpatch**

The *E*- and *H*-plane patterns are stored in separate files with extensions of .epl and .hpl, respectively. Two file formats are available. In one, the data is stored as angle and field (in decibels) value pairs separated by a comma. With the other, there are three blank spaces and no comma between the angle and field.

INSERT COMMAS BETWEEN ANGLE AND PATTERN VALUE (y or n)? **n**

Finally the user can either do another case or exit.

DO ANOTHER CASE (y or n)? **n**

3.8 QUARTERWAVE SHORT-CIRCUITED PATCH

The dominant electric field component underneath the patch is oriented vertically, that is between the patch and the ground plane. Its distribution with length has an approximate cosinusoidal variation. The maximum value is at the input edge, goes to zero at the center, and returns to a maximum at the other edge. Since the *E*-field is zero there, a short circuit

(a)

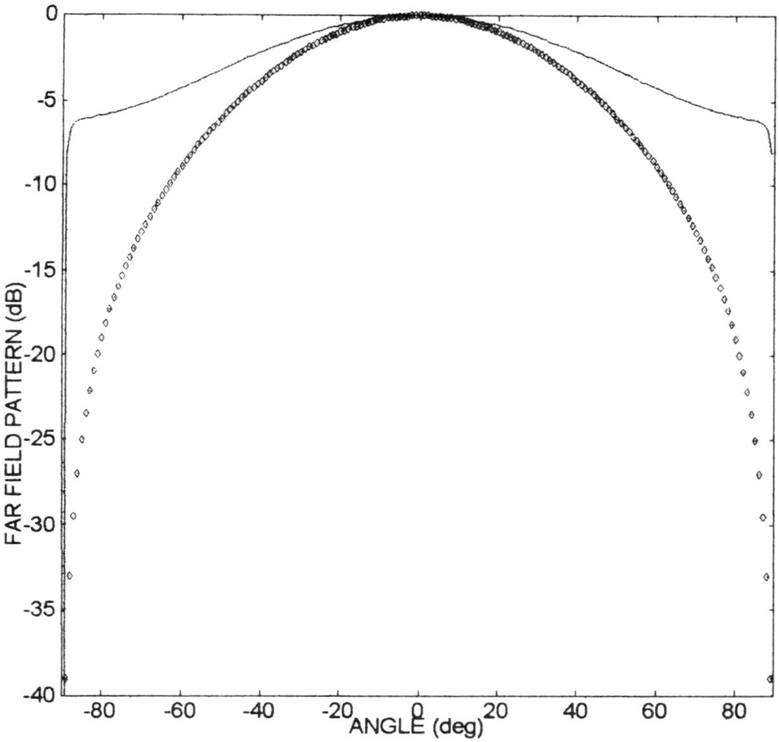

Figure 3.6 Screen display of patch radiation pattern: (a) rectangular plot and (b) polar plot.

can be placed at the center, and basic operation will not be affected. The nominal length becomes about a quarter wavelength. This type of patch is referred to as a short circuit patch (Figure 3.7).

Short-circuited patches are used in situations where there is not enough room for a standard patch. They are also used when a broader *E*-plane pattern is needed. The short-circuited patch has only one radiating edge. The *E*-plane pattern is broader because no array with the second edge is formed. There is increased cross-polarized radiation as well. Being smaller than a normal patch, the short-circuited patch bandwidth is about 80% that of the corresponding full-sized antenna. Also the input impedance at resonance is about 2.5 times higher [15].

A common technique for realizing the short circuit is to use shorting pins or plated through via holes. These appear as an inductive reactance rather than a pure short. The patch length must be adjusted to account for this added inductance. A model can be derived for the shorting pins by assuming the pins are across a parallel plate transmission

(b)

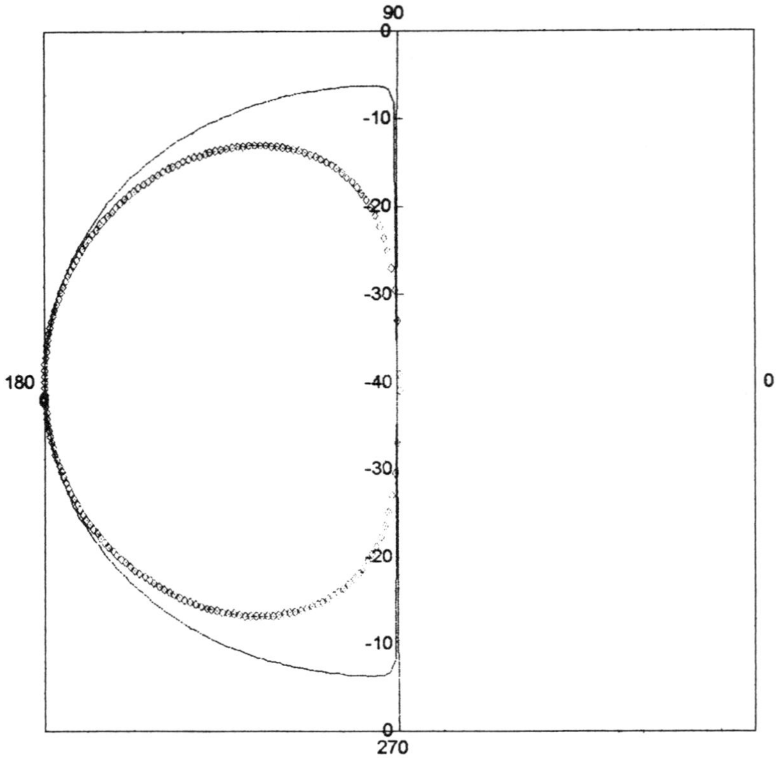

Figure 3.6 (continued)

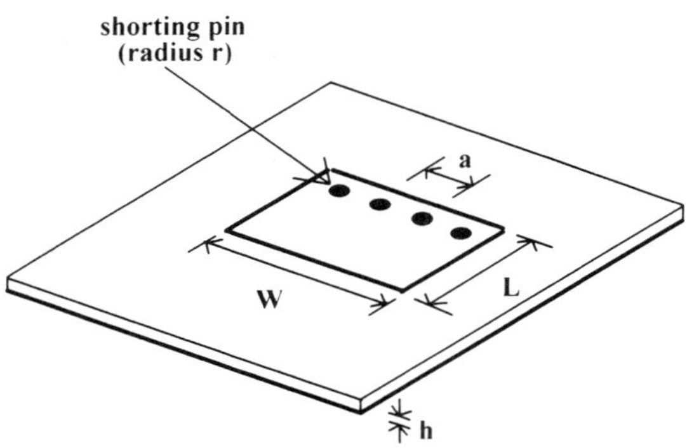

Figure 3.7 Quarter-wavelength-long patch with shorting pins.

line backed by an open circuit. The equivalent circuit elements for this model are then used to determine end correction for the patch length. The change in patch length is given by [15]

$$\frac{\Delta l}{a} = \frac{1}{2\pi}\left\{\ln\left(\frac{a}{2\pi r}\right) - 4\ \pi^2\frac{r^2}{a^2} + 0.601\frac{a^2}{\lambda^2}\right\} \tag{3.17}$$

where a is the separation between pins, r the pin radius, and λ the wavelength in the dielectric.

SCPATCH uses a transmission line model to analyze and design quarterwave patches. The model is the same as that in PATCH9 with one radiating edge and the mutual impedance between edges deleted. SCPATCH runs in exactly like PATCH9 with the only difference being the inputs for shorting pins, if used, via the prompt

Use shorting pins for short circuit (y or n)? **y**

If answered no, the program assumes a perfect short. When answered yes, the following inputs are requested:

Input shorting pin diameter (cm)? **.0635**
(2r in Figure 3.7)

Input center-to-center spacing between pins (cm)? **.254**

(a in Figure 3.7)

The 1.575-GHz example of Section 3.3 is redesigned to be a short-circuited patch using SCPATCH. The results are as follows.

SCPATCH.V10 11-04-1995 18:01:58
Current Program Date - 16/03/1995

Substrate height (cm) = 0.160
Relative dielectric constant = 2.200
Loss tangent = 0.0010
Line thickness (cm) = 0.00356
Patch width (cm) = 9.434
Patch length (cm) = 3.153
Feed type = microstrip line
Feed point inset (cm) = 0.0000
Feed line width (cm) = 0.495
Shorting pin dia = 0.0635 (cm) Pin spacing = 0.254 (cm)
Resonant frequency (GHz) = 1.575000
Resonant impedance = 368.201 j −0.0251 ohms

FREQUENCY (GHz)	*INPUT IMPEDANCE (ohms)*
1.550	*59.748 J 137.729*
1.555	*85.990 J 157.755*
1.560	*130.001 J 177.949*
1.565	*203.802 J 185.023*
1.570	*307.313 J 138.775*
1.575	*368.201 J −0.025*
1.580	*305.373 J −136.551*
1.585	*203.535 J −181.104*
1.590	*131.329 J −174.404*
1.595	*88.012 J −155.063*
1.600	*61.956 J −135.773*

Notice the increased resonant impedance and reduced bandwidth as compared to the full patch. SCPATCH also has an analysis mode.

3.9 PATCH WITH COVER LAYER

It may be necessary to provide some environmental protection for the microstrip patch. One way is to use a radome consisting of a dielectric sheet located some distance away from the antenna. To minimize the radome's impact, make the patch to radome separation as large as possible, preferably a wavelength or more. The radome material should have a low dielectric constant and loss tangent. For planar radomes, the optimum thickness is a half-wavelength or some multiple thereof. When utilized in this manner, the radome may have some effect on the pattern, but the patch resonant frequency and impedance remain essentially unchanged.

Another patch protection technique is to place a cover layer directly over the patch substrate. (See Figure 3.8.) The cover could be another piece of substrate material. This makes for a very compact structure since there is no gap between the patch and its cover. The patch properties will change, however, because the cover interacts with the fringing fields around the patch. The resonant frequency shifts downward, and the input impedance is modified. The cover can have a profound effect on the radiation patterns as well.

A modified transmission line model [16] can be used to analyze the covered patch. The characteristic impedance, effective dielectric constant, and losses of the transmission line are those of a covered microstrip. The slot admittance becomes altered because of the additional layer. The radiating slot capacitance is found from

$$C_s = \frac{W}{2}\left[\frac{\epsilon_{ra}(L)}{cZ_0(L)} - \epsilon_0\epsilon_r\frac{L}{h_s}\right] \tag{3.18}$$

where $Z_0(L)$ and $\epsilon_{ra}(L)$ are the characteristic impedance and effective dielectric constants for a microstrip line of width L. Equation (3.18) is the edge capacitance of a covered

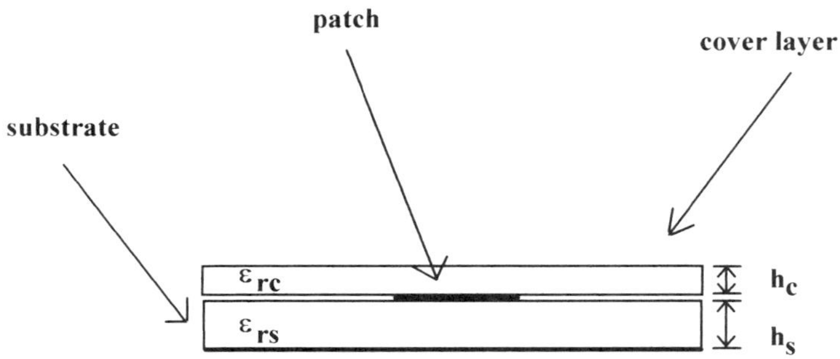

Figure 3.8 Patch with protective cover layer.

microstrip line of width L instead of W. The first term is the actual line capacitance, while the second is the capacitance per unit length of a parallel plate capacitor of width L. The difference is the fringing field capacitance. This serves as a reasonable approximation to the slot capacitance. The change in slot conductance is more difficult to determine. Reference [16] gives integrals for the slot conductance caused by radiation and surface waves. These are not used here because their numerical evaluation is beyond the intended scope of the models. The slot conductance and mutual admittance between slots are evaluated as in the uncovered model [3]. Fortunately the change in conductance has less impact than with the capacitance. Comparison with measurements will show that this is not a serious omission.

The parameters of the covered microstrip are derived using a variational technique [17]. The line is assumed to support a quasi-TEM mode. For a TEM mode, the characteristic impedance and effective dielectric constant are

$$Z_0 = \frac{1}{c\sqrt{C_0 C}} \tag{3.19}$$

$$\epsilon_{\text{eff}} = \frac{C}{C_0} \tag{3.20}$$

where C and C_0 are the capacitances of the structure with and without the dielectrics. Poisson's equation subject to the appropriate boundary conditions is solved in the Fourier transform domain. Applying the variational method yields an integral equation for the capacitance in terms of the unknown charge distribution on the microstrip. The nature of the variational method is such that an approximate function for the distribution can be used with only a second-order error in the resulting calculation. The expression derived by Bahl and Stuchly [17] for the capacitance is

$$\frac{1}{C} = \frac{1}{2\pi Q^2}\int \frac{f^2(\beta)d(\beta h_s)}{[\epsilon_{rc}[\epsilon_{rc}\tanh(\beta h_c) + 1]/[\epsilon_{rc} + \tanh(\beta h_c)] + \epsilon_{rs}\tanh(\beta h_s)](\beta h_s)} \tag{3.21}$$

where

$$\frac{f(\beta)}{Q} = 1.6\left\{\frac{\sin(\beta W/2)}{BW/2}\right\} + \frac{2.4}{(\beta W/2)^2}\left\{\cos\left(\frac{\beta W}{2}\right) - \frac{2\sin(\beta W/2)}{\beta W/2} + \frac{\sin^2(BW/4)}{(\beta W/4)^2}\right\} \tag{3.22}$$

is the Fourier transform of the charge distribution approximating function. The integral in (3.21) is easily evaluated numerically.

The above analysis has been implemented in PATCHC, which designs dielectric covered patches. Its results have been compared to the experimental data presented in [18]. Some experimental details like the type and position of the feed were not stated in the article. The uncovered patch was resonant at 4.1 GHz. The substrate had an ϵ_r of 2.32 and a thickness of 0.159 cm, which is fairly thick ($\approx 0.02\lambda_0$). Five different cover layers were studied with ϵ_r's from 2.32 to 10.2 and thickness from 0.00508 cm to 0.635 cm. PATCHC was used in both the design and analysis modes to compare against the measured results.

In the analysis mode, the patch length used in [18] was held constant and the frequency varied until resonance was obtained. For the lower dielectric constant covers, the calculated resonant frequency was within 2% of the experimental value, usually higher than measured. With very thin cover layers (≤ 0.0127 cm), PATCHC predicted a slightly higher frequency (still within 2%) than for the measured uncovered patch. For the high ϵ_r substrates, the calculated values were between 3% and 5% higher compared to measurements. In the design mode, the measured frequency was input, and PATCHC asked to find the resonant length. The calculated lengths were within the same accuracies as the predicted frequencies. They were always longer than the actual length. Given the relatively thick substrate (and in most cases, covers), the agreement is reasonable. For design purposes, try designing the patch about 2% higher in frequency since the calculated lengths tend to be too long.

PATCHC is identical to PATCH9 with the only difference being the need to input the height and electrical parameters of the cover layer. A sample output from one of the examples in [18] is as follows.

PATCHC.V12 05-25-1995 22:23:17
Current Program Date - 16/03/1995

Substrate height (cm) = 0.159
Substrate relative dielectric constant = 2.320
Substrate loss tangent = 0.0010
Line thickness (cm) = 0.00356
Cover layer height (cm) = 0.3175

Cover layer relative dielectric constant = 2.600
Cover layer loss tangent = 0.0010
Patch width (cm) = 1.900
Patch length (cm) = 2.318
Feed point inset (cm) = 0.0000
Resonant frequency (GHz) = 3.874000
Resonant impedance = 386.538 j 0.0482 ohms

FREQUENCY (GHz)	INPUT IMPEDANCE (ohms)
3.774	82.359 J 159.209
3.794	115.581 J 177.950
3.814	167.528 J 192.582
3.834	245.367 J 187.201
3.854	338.564 J 128.585
3.874	386.538 J 0.048
3.894	337.831 J −127.041
3.914	246.447 J −184.522
3.934	170.482 J −190.583
3.954	119.464 J −177.233
3.974	86.505 J −159.664

PATCHC also can be used for analysis.

Cover layers directly on top of the patch can produce changes in the radiation pattern. If pattern shape is important, the effect of the cover should be ascertained. The approach in [19] extends that of [7] to include a cover layer. As in [7], the field at the patch location due to a distant dipole is found. Being far away, the fields look essentially like plane waves. With plane-wave incidence, the dielectric layers can be treated as transmission lines, and it is relatively straightforward to determine the fields at the patch location. By reciprocity, this field corresponds to that radiated by a current element on the patch to the dipole. Superposition accounts for the actual current distribution on the patch.

Following the development in Chapter 2, the radiated field becomes

$$E_\theta \propto -\cos(\phi)\overline{J_{sx}}(k_x, k_y)G(\theta) \tag{3.23}$$

$$E_\phi \propto \sin(\phi)\overline{J_{sx}}(k_x, k_y)F(\theta) \tag{3.24}$$

where $\overline{J_{sx}}(k_x, k_y)$ is the Fourier transform of the patch current density as given in Chapter 2. The factors $G(\theta)$ and $F(\theta)$ account for the two dielectric layers

$$G(\theta) = 2\tan[\beta_1 h_s]\sec[\beta_2 h_c] \cdot \left[\tan[\beta_1 h_s] + \frac{\epsilon_{rs} n_1(\theta)}{\epsilon_{rc} n_2(\theta)}\tan[\beta_2 h_c] - j\left\{\frac{\epsilon_{rs}}{n_1(\theta)}\cos(\theta)(1 - \frac{\epsilon_{rc} n_1(\theta)}{\epsilon_{rs} n_2(\theta)}\tan[\beta_1 h_s]\tan[\beta_2 h_c])\right\}\right]^{-1} \tag{3.25}$$

$$F(\theta) = 2\tan[\beta_1 h_s]\sec[\beta_2 h_c] \cdot \left[\tan[\beta_1 h_s] + \frac{\mu_{rc} n_1(\theta)}{\mu_{rs} n_2(\theta)}\tan[\beta_2 h_c] - j\left\{\frac{n_1(\theta)}{\mu_{rs}}\sec(\theta)(1 - \frac{\mu_{rs} n_2(\theta)}{\mu_{rc} n_1(\theta)}\tan[\beta_1 h_s]\tan[\beta_2 h_c])\right\}\right]^{-1} \tag{3.26}$$

with

$$\beta_1 = k_0 n_1(\theta) \qquad \beta_2 = k_0 n_2(\theta)$$

and

$$n_1(\theta) = \sqrt{n_1^2 - \sin^2(\theta)} \qquad n_2(\theta) = \sqrt{n^2 - \sin^2(\theta)}$$

where $n_1 = \sqrt{\epsilon_{rs}\epsilon_0\mu_{rs}\mu_0}$, $n_2 = \sqrt{\epsilon_{rc}\epsilon_0\mu_{rc}\mu_0}$, and μ_r is the substrate relative permeability (usually 1). The angle ϕ is the azimuth angle in the plane of the patch. θ is the elevation angle with 0 degrees being broadside to the patch. The subscripts s and c refer to the substrate and cover layers respectively. These equations have been programmed in CHWPATCH.

CHWPATCH requires the usual pattern calculation inputs (for example, substrate and cover layer parameters, patch dimensions, frequency, and pattern angles). Radiation patterns may be displayed in either a rectangular or polar format. They can also be stored in files, one for the *E*-plane (with extension .epl) and another for the *H*-plane (extension .hpl).

The pattern for the 3.874-GHz antenna of the previous example is given in Figure 3.9. Note the narrow *H*-plane pattern even though the patch width is less than the length. The *E*-plane shows slight changes as well.

3.10 CIRCULAR PATCH DESIGN

Another commonly used patch shape is the circular disk (see Figure 3.10). Circular patches tend to be slightly smaller than rectangular patches and thus could be candidates when

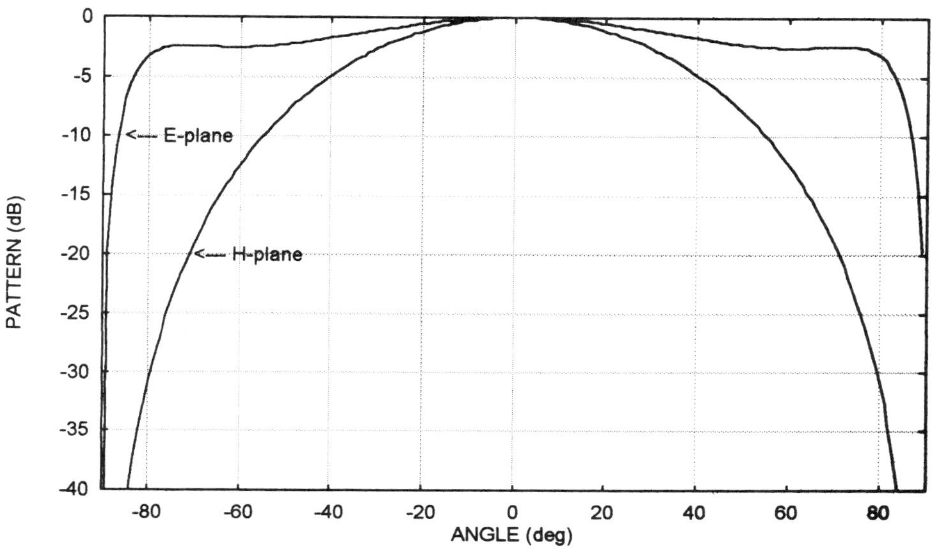

Figure 3.9 Radiation patterns for a covered 3.874-GHz patch.

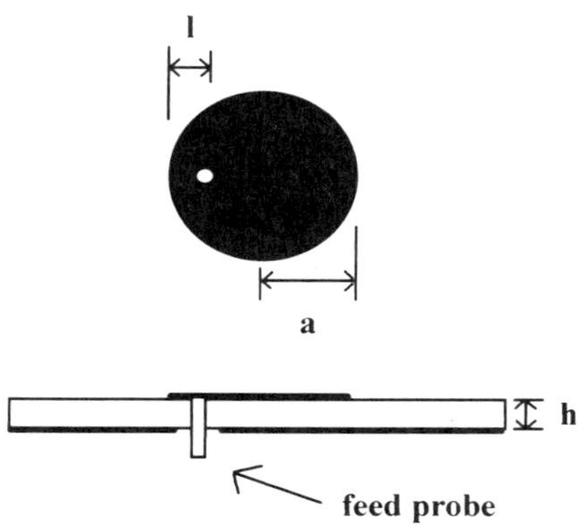

Figure 3.10 Circular patch geometry.

space is limited. Circular patches also lack a preferred axis (for example, there is no length or width). The feed can be placed anywhere instead of just along the length.

Patch design consists of substrate selection, feed type choice, and feed location determination. The radius is determined by the condition for resonance. The approximate equation of resonance is $J_n'(\sqrt{\epsilon_r} k_0 a) = 0$ where J_n is the Bessel function of order n, and a is the radius. The prime denotes differentiation. For the lowest order mode, $n = 1$, the first root of the equation equals 1.841. The input impedance can be varied by insetting the feed. As the feed inset increases toward the center, the impedance lowers.

Both microstrip and probe feeds are suitable for circular patches. Microstrip feeds are not inset since this would destroy the patch symmetry. The dominant mode has a cosine E-field distribution with azimuth angle around the patch. The field is maximum at the feed, so any removal of patch metal would cause different modes to be excited. Probe feeds are usually inset because the circular patch has a high impedance.

The impact of several design parameters is shown in Table 3.3. These were determined by using PATCHD. A patch for 1.575-GHz was analyzed. In general, the circular patch has a smaller bandwidth, lower efficiency, and higher input resistance than a rectangular patch. This is a result of a smaller antenna size. The changes in patch performance with substrate parameters follow those of the rectangular case.

Circular patches can be designed with program CPATCH. It is not a transmission line model. Instead it is based on the curve-fit equations of PATCHD. Although a generalized transmission line model is available [20], it has not seen widespread application except for rectangular shapes. Transmission line models tend to get more complicated for nonrectangular geometries. CPATCH finds the resonant radius using the curve-fit equations. The resonant resistance is [21]

$$R_{\mathrm{T}} = \frac{Q_{\mathrm{total}}}{\omega C_{\mathrm{e}}} J_1^2\left(\frac{1.841\,\omega_{\mathrm{r}}(a - l)}{a}\right) \tag{3.27}$$

where $C_{\mathrm{e}} = \dfrac{\epsilon_r \epsilon_0 a^2}{2.775h}$, ω_{r} is the real part of the resonant radian frequency, J_1 is the Bessel function of order 1, and l is the feed inset (probe feed only).

Near the resonant frequency, the patch is modeled as a parallel RLC circuit. The input impedance becomes [9]

Table 3.3
Circular Patch Characteristics Versus Design Variables

Substrate ϵ_r	*Substrate Height (cm)*	*Patch Radius (cm)*	*Input Resis. (Ohms)*	*Radiation Efficiency (%)*	*Total Efficiency (%)*	*Bandwidth (%)*
2.2	0.16	3.675	369.5	96.9	84.2	1.10
2.2	0.0635	3.726	249.1	98.7	58.8	0.63
2.2	0.635	3.444	443.6	88.7	86.8	4.03
1.1	0.16	5.164	344.0	99.9	87.9	1.20
4.4	0.16	2.634	525.1	93.2	75.2	0.75

$$Z_{in} = \frac{R_T[1 - Q_{total}(f/f_r - f_r/f)]}{1 + Q_{total}^2(f/f_r - f_r/f)^2} \quad (3.28)$$

where f_r is the resonant frequency. Because CPATCH is not a transmission line model, the microstrip feed cannot be easily included. There is not a convenient way to reduce the radiating edge conductance by the shadowing of the line. A probe feed model, identical to that in PATCH9, is included.

When run, CPATCH asks for frequency, substrate height, relative ϵ_r, tan δ, and relative conductivity. It calculates the radius and complex resonant frequency. From these, it can find such values as R_T and Q_{TOT}. The user is then asked if a probe feed is used. For probe feeds, the probe and relief hole diameters are input and the equivalent circuit elements calculated. Finally, frequency data for impedance calculation (for example, start and stop frequencies plus number of frequencies) are requested. The program then determines the impedance versus frequency. The results for a 1.575-GHz patch are as follows.

CPATCH.V10 11-04-1995 18:25:05

Circular Patch Design

SUBSTRATE HEIGHT = 0.160 cm
SUBSTRATE RELATIVE DIELECTRIC CONSTANT = 2.20
SUBSTRATE LOSS TANGENT = 0.0010
CONDUCTOR RELATIVE CONDUCTIVITY = 1.000
PATCH RADIUS = 3.675 cm
FEED LOCATION = 3.556 cm
FREQUENCY = 1.5742 GHz

FREQ (GHz)	*PATCH IMPEDANCE (ohms)*
1.550	74.52 j 147.76
1.555	106.04 j 166.52
1.560	156.85 j 181.78
1.565	235.47 j 176.34
1.570	329.41 j 112.05
1.575	365.87 j −24.60
1.580	300.24 j −142.13
1.585	207.49 j −182.22
1.590	139.06 j −178.24
1.595	95.80 j −161.34
1.600	68.72 j −143.30

The computations of CPATCH have been compared with experimental data. Data for a circular patch on a substrate with $\epsilon_r = 2.47$ is given in [22]. Three substrate heights,

h/λ_0 = 0.0034, 0.0071, and 0.0149 were tested. CPATCH predicts the patch radius to within 2%, with the computed value being higher. For the two thicker substrates, the calculated resonant resistance is about 10% to 12% too high. Strangely enough, with the thinnest substrate, the CPATCH resistance is about 65% too high. The measured patches are probe fed, but no probe details are given, so their effect is unknown. Except for these very thin substrates, the accuracy is typical for closed-form models. CPATCH does not have an analysis mode.

CIRPAT calculates radiation patterns for circular patches. It is based on the analysis presented in [23]. The cavity model is used to find the fields at the patch edge. These are converted into equivalent electric and magnetic current sources. Initially the substrate is ignored and the radiated fields are determined by standard aperture antenna theory according to the relations

$$E_\theta = -j\frac{e^{-jk_0 r}}{r}\frac{aE_0k_0h}{2}[J_0(k_0a\sin(\theta)) - J_2(k_0a\sin(\theta))]\cos(\phi) \tag{3.29}$$

$$E_\phi = -j\frac{e^{-jk_0 r}}{r}\frac{aE_0k_0h}{2}[J_0(k_0a\sin(\theta)) + J_2(k_0a\sin(\theta))]\sin(\phi)\cos(\theta) \tag{3.30}$$

$J_0(\ )$ and $J_2(\ )$ are Bessel functions of orders 0 and 2, respectively. θ and ϕ have the same meanings as before. Factors—derived from a rigorous solution—that account for the dielectric are added. The factor that multiplies (3.29) is

$$F_\theta = \frac{\cos(k_{1z}d)[1 + T^2\tan^2(k_{1z}d)]}{1 - \tan^2(k_{1z}d) + j\dfrac{2\tan(k_{1z}d)}{T}} \tag{3.31}$$

while (3.30) is multiplied by

$$F_\phi = \frac{\cos(k_{1z}d)\left[1 + \dfrac{\epsilon_r^2\tan^2(k_{1z}d)}{T^2}\right]}{1 - \tan^2(k_{1z}d) + j\dfrac{2\tan(k_{1z}d)}{\epsilon_r}} \tag{3.32}$$

where $k_{1z} = k_0\sqrt{\epsilon_r - \sin^2(\theta)}$, $d = h/2$, and $T = \epsilon_r\cos(\theta)\sqrt{\epsilon_r - \sin^2(\theta)}$. Patterns for the above patch are shown in Figure 3.11. Compared to a similar rectangular antenna, the circular one has a narrower E-plane and wider H-plane patterns. CIRPAT is used in the same manner as the other pattern programs.

3.11 WRAPAROUND ELEMENT DESIGN

Although the rectangular and circular patches are the most commonly used, one of the earliest microstrip patch antennas was the wraparound patch [24]. It consists of a very

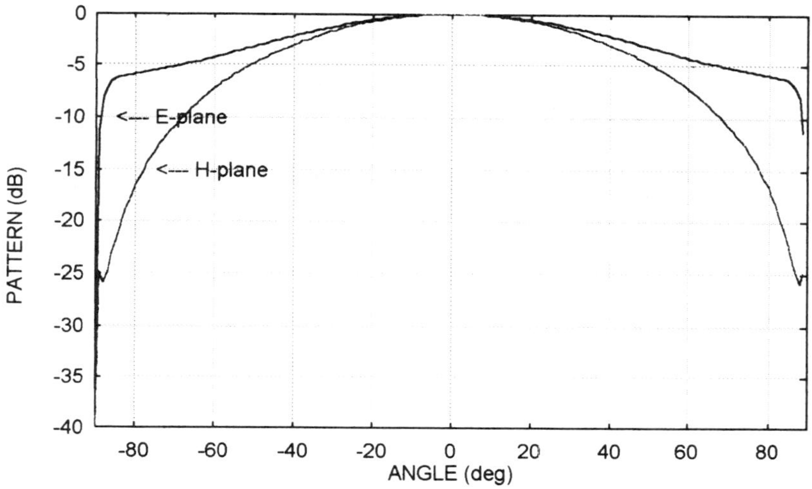

Figure 3.11 *E*- and *H*-plane patterns for a 3.863-cm-radius circular patch at 1.575 GHz.

long rectangular patch that is completely wrapped around a cylinder as in Figure 3.12. The patch forms a belt that totally surrounds the cylinder. Thin, flexible, PTFE-based substrates are generally used for this application. Multiple feed points are required, usually in conjunction with a parallel (corporate) type feed. These antennas found use for telemetry and communications on missiles and rockets.

The design of a wraparound consists of substrate selection, resonant length calculation, determination of number of feeds, calculation of feed point input impedance, and design of the corporate feed. Figure 3.13 shows a flattened out wraparound antenna. The

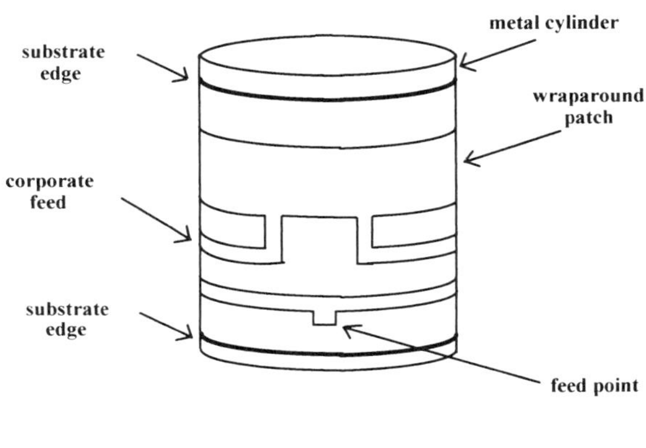

Figure 3.12 Wraparound patch antenna.

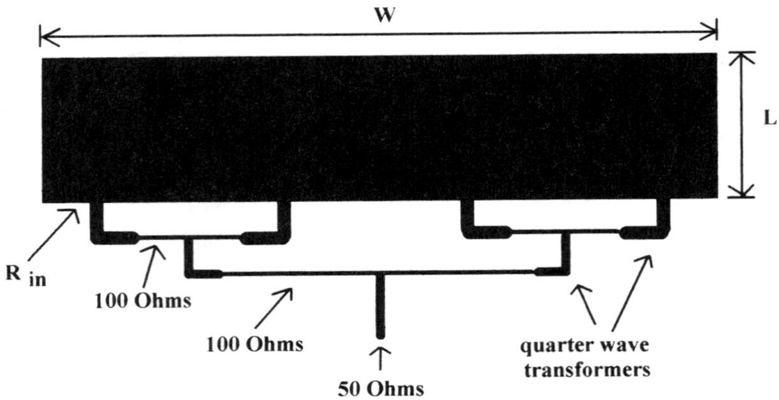

Figure 3.13 Wraparound with feed network.

resonant length criterion is the same as for a rectangular patch; that is, it is the length that makes the input impedance real. The length is found from the dielectric constant of the microstrip line, which has a width $W = \pi D$, where D is the cylinder diameter. Since the patch is closed upon itself, there are no fringing fields and the effective dielectric constant is essentially the dielectric constant of the substrate. Calculation of the wavelength in the patch follows. The length is chosen to be one half-wavelength minus twice the end-effect extension.

The number of feeds is related to the width, W. The field under the patch must be uniform across the width. When the width gets larger than a wavelength, higher order transverse modes with nonuniform fields can be excited. To prevent this, the number of feeds must equal or exceed the number of wavelengths that the patch is wide [24]. The width of the patch in wavelengths is

$$W_D = \frac{W\sqrt{\epsilon_{\text{eff}}}}{\lambda_0} \tag{3.33}$$

The number of feeds, N_F, must exceed W_D. The most common feed network is a parallel (corporate) feed consisting of two-way power splitters and equal length lines as shown in Figure 3.12. With this type of feed, N_F must be a power of two as well as be greater than W_D.

Munson [24] found that the radiating slot conductance and susceptance could be approximated by expressions valid for a narrow slot in an infinite ground plane. On a per unit width basis, they are

$$G_s = \frac{\pi}{\lambda\eta}\left[1 - \frac{(kh)^2}{24}\right] \tag{3.34}$$

$$B_s = \frac{3.135 - 2\log(kh)}{\lambda\eta} \tag{3.35}$$

where $\eta = 377\Omega$. To get the admittance for the slot, (3.33) and (3.34) must be multiplied by W. At resonance, the susceptance is canceled out, leaving only the conductance of the two slots in parallel. Thus the resonant resistance is $R_{in} = 1/2WG_s$. This is the total patch resistance. Since the feeds are in parallel, each feed sees a value of $R_f = N_F R_{in}$.

The antenna resistance at each feed is transformed to some convenient value, usually 100Ω by a quarterwave transformer. The quarterwave transformer impedance is $Z_t = \sqrt{100R_{in}}$ (see Figure 3.12). The next feed section is a 100-Ω line. Two equal length lines from adjacent feeds meet at the splitter junction. The two feed sections are in parallel, resulting in an impedance of 50Ω. At the junction, another quarterwave transformer of impedance 70.7Ω brings the impedance level back to 100Ω. The process is repeated until the final splitter, where the output is a 50-Ω line of arbitrary length. It is important to maintain symmetry and equal length line sections to each feed. This ensures an in-phase excitation.

In some cases it may not be possible to transform the feed impedance to 100Ω. The required transformer line width may be too narrow or too wide. As a rule, microstrip impedances should be between 20Ω and 110Ω. The narrowest line width depends upon the fabrication accuracy. With narrow widths, small errors produce large changes in impedance. Very wide lengths may suffer from transverse mode excitation. For a given width, the cut-off frequency for the lowest transverse mode is [25]

$$f_{CT} = \frac{c}{\sqrt{\epsilon_r}(2W + 0.8h)} \tag{3.36}$$

This frequency should be well above the maximum operating frequency. Wide lines also can cause blockage of the patch radiating edge. If the transformer is not realizable, transform the patch impedance to a lower value with a realizable width line. Then transform the impedance back up to 100Ω.

The wraparound patch model of [24] has been incorporated into the program WRAPPAT. WRAPPAT determines the resonant length, the number of feeds, and the impedance at each feed. The inputs are substrate parameters, cylinder diameter, and frequency. A sample output is

```
WRAPPAT.V10      11-04-1995      18:29:07
            Wraparound Patch Design

SUBSTRATE HEIGHT = 0.0795 (cm)
SUBSTRATE er = 2.20 AND loss tangent = 0.0010
CONDUCTOR THICKNESS = 0.00356 AND relative conductivity = 1.0000
CYLINDER DIAMETER = 29.210 (cm)
DESIGN FREQUENCY = 1.5750 GHz
```

PATCH LENGTH = 6.312 (cm)
NUMBER OF FEEDS = 8
IMPEDANCE PER FEED = 99.64 (Ohms)

WRAPPAT does not compute the impedance versus frequency. Although the transmission line model gives reasonable results for the resonant length, it does not work well for the frequency behavior of the impedance. This may be due to the radiator being closed upon itself.

Unlike rectangular patches, calculation of wraparound radiation patterns is a nontrivial problem. Ashkenazy et al. [26] derived an expression by solving the wave equation (in cylindrical coordinates) in the Fourier domain subject to the appropriate boundary conditions and an assumed patch surface current distribution. With this analysis, the far field pattern is

$$f_{wr}(\theta) = \frac{\cos((\pi \cos\theta)/2\sqrt{\epsilon_r})}{\epsilon_r - \cos^2\theta} \left| \frac{\epsilon_r \sin\theta}{\sqrt{\epsilon_r - \cos^2\theta}} Z(\theta) H_0^{(2)}(k_0 b \sin\theta) - H_1^{(2)}(k_0 b \sin\theta) \right|^{-1} \quad (3.37)$$

with

$$Z(\theta) = \frac{H_1^{(1)}(x_b)H_0^{(2)}(x_a) - H_0^{(1)}(x_a)H_1^{(2)}(x_b)}{H_0^{(1)}(x_b)H_0^{(2)}(x_a) - H_0^{(1)}(x_a)H_0^{(2)}(x_b)}$$

$$x_a = k_0 a\sqrt{\epsilon_r - \cos^2\theta}, \qquad x_b = k_0 b\sqrt{\epsilon_r - \cos^2\theta}$$

where a is the cylinder radius, b is the radius plus the substrate thickness, $H_{0,1}{}^{(1,\,2)}(x_{a,\,b})$ are Hankel functions of the first kind, order 0, and the second kind, order 2, and θ is the pattern angle measured from the cylinder axis (90 degrees and 270 degrees are broadside to the cylinder). These equations have been programmed in WRAPRND.

Figure 3.14 shows the pattern calculated by WRAPRND for the antenna just analyzed. The cut shown is through the cylinder axis. The cylinder lies along the 0-degrees to 180-degrees line in the figure. The pattern is omnidirectional in the plane perpendicular to the cylinder.

The factors that affect the pattern are cylinder diameter, substrate height, and dielectric constant. For small diameters, thin substrates, and very low dielectric constants ($\epsilon_r \rightarrow 1$), the pattern becomes like that of a dipole. The pattern maximum is perpendicular to the axis and has a 3-dB beamwidth of about 60 degrees. As any of those parameters increases, the pattern broadens out, becoming more and more omnidirectional, as seen in Figure 3.13.

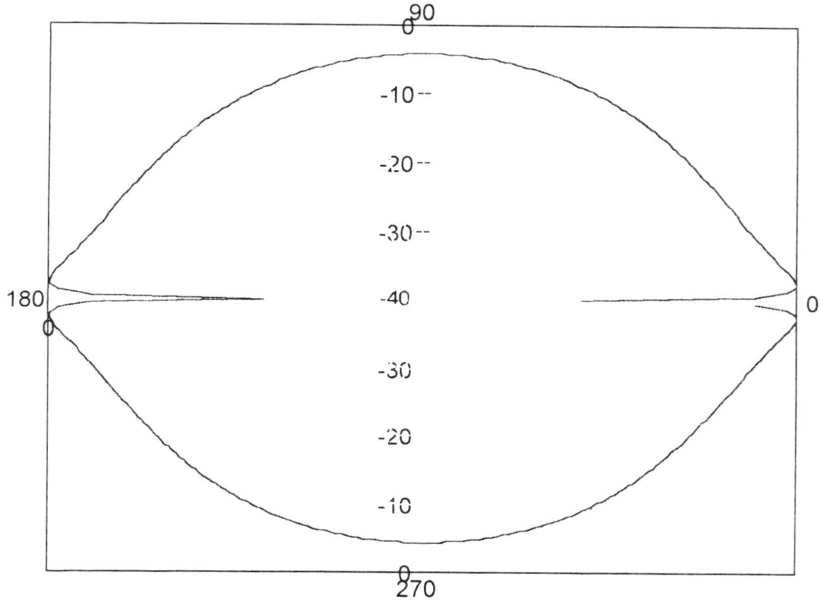

Figure 3.14 Wraparound patch pattern through the cylinder axis.

References

[1] James, J. R., A. Henderson, and P. S. Hall, "Microstrip Antenna Performance Is Determined by Substrate Constraints," *MSN*, Aug. 1982, pp. 73–84.

[2] Bhartia, P., K. V. S. Rao, and R. S. Tomar, *Millimeter-Wave Microstrip and Printed Circuit Antennas*, Norwood, MA: Artech House, 1991, pp. 58–59.

[3] Pues, H., and A. Van de Capelle, "Accurate Transmission Line Model for the Rectangular Microstrip Antenna," *PROC IEE, Pt. H (Microwaves, Optics, & Acoustics)*, Vol. 131, No. 6, 1984, pp. 334–340.

[4] Chew, W. C., and Q. Liu, "Resonance Frequency of a Rectangular Microstrip Patch," *IEEE Trans. on Antennas and Propagation*, Vol. 36, No. 8, Aug. 1988, pp. 1045–1056 (see also corrections, *IEEE Trans. on Antennas and Propagation*, Vol. 12, No. 12, Dec. 1988, p. 1827).

[5] Liu, Q., and W. C. Chew, "Curve-Fitting Formulas for Fast Determination of Accurate Resonant Frequency of Circular Microstrip Patches," *Proc. IEE Pt. H (Microwaves, Optics and Acoustics)*, Vol. 135, No. 5, Oct. 1988, pp. 289–292.

[6] Pozar, D. M., "Rigorous Closed-Form Expressions for the Surface Wave Loss of Printed Circuit Antennas," *Electron Lett.*, Vol. 26, No. 13, June 1990, pp. 954–955.

[7] Jackson, D., and N. Alexopoulos, "Simple Formulas for the Input Impedance, Bandwidth, and Radiation Efficiency of a Rectangular Patch," *IEEE Int. Symp. Digest Antennas and Propagation*, San Jose, CA,1989, pp. 1130–1133.

[8] Pues, H., and A. Van de Capelle, "An Impedance-Matching Technique for Increasing the Bandwidth of Microstrip Antennas," *IEEE Trans. on Antennas and Propagation*, Vol. 37, No. 11, Nov. 1989, pp. 1345–1354.

[9] Abboud, F., J. P. Damiano, and A. Papiernik, ''Simple Model for the Input Impedance of Coax-Fed Rectangular Microstrip Patch Antenna for CAD,'' *IEE Proc. Pt. H (Microwaves, Optics and Acoustics)*, Vol. 135, No. 3, Oct. 1988, pp. 323–326.

[10] Kirshning, M., R. H. Jansen, and N. H. L. Koster, ''Accurate Model for Open-Ended Effect of Microstrip Lines,'' *Electronic Lett.*, Vol. 17, No. 3, Feb. 1981, pp. 123–125.

[11] Zheng, J.-X., and D. C. Chang, ''End-Correction Network of a Coaxial Probe for Microstrip Patch Antennas,'' *IEEE Trans. on Antennas and Propagation*, Vol. 39, No. 1, Jan. 1991, pp. 115–118.

[12] Schaubert, D. E., D. M. Pozar, and A. Adrian, ''Effect of Microstrip Antenna Substrate Thickness and Permittivity: Comparison of Theories with Experiment,'' *IEEE Trans. on Antennas and Propagation*, Vol. 17, No. 6, June 1989, pp. 677–682.

[13] Huang, J., ''The Finite Ground Plane Effect on the Microstrip Antenna Radiation Pattern,'' *IEEE Trans. on Antennas and Propagation*, Vol. 31, No. 4, July 1983, pp. 649–653.

[14] Balanis, C. A., *Advanced Engineering Electromagnetics*. New York, NY: John Wiley & Sons, 1989, Chapter 13.

[15] James, J. R., P. S. Hall, and C. Wood, *Microstrip Antenna, Theory and Design*, London, UK: Peter Peregrinus, 1981, pp. 102–106.

[16] Shavit, R., ''Dielectric Cover Effect on Rectangular Microstrip Antenna Array,'' *IEEE Trans. on Antennas and Propagation*, Vol. 42, No. 8, Aug. 1994, pp. 1180–1184.

[17] Bahl, I. J., and S. S. Stuchly, ''Analysis of a Microstrip Covered With a Lossy Dielectric,'' *IEEE Trans. on Microwave Theory and Techniques*, Vol. 28, No. 2, Feb. 1980, pp. 104–109.

[18] Bahl, I. J., P. Bhartia, and S. S. Stuchly, ''Design of Microstrip Antennas Covered with a Dielectric Layer,'' *IEEE Trans. on Antennas and Propagation*, Vol. 30, No. 2, Feb. 1982, pp. 314–318.

[19] Jackson, D. R., and N. G. Alexopoulos, ''Gain Enhancement Methods for Printed Circuit Antennas,'' *IEEE Trans. on Antennas and Propagation*, Vol. 33, No. 9, Sept. 1985, pp. 976–987.

[20] Bhattacharyya, A. K., and R. Garg, ''Generalised Transmission Line Model for Microstrip Patches,'' *IEE Proc. Pt. H (Microwaves, Optics and Acoustics)*, Vol. 132, No. 1, 1985, pp. 93–98.

[21] Bahl, I. J., and P. Bhartia, *Microstrip Antennas*, Norwood, MA: Artech House, 1980, pp. 88–105.

[22] Long, S. A., L. C. Shen, M. D. Walton, and M. R. Allerding, ''Impedance of a Circular-Disc Printed-Circuit Antenna,'' *Electronic Lett.*, Vol. 14, No. 21, Oct. 1978, pp. 684–686.

[23] Shafai, L., and A. K. Bhattacharyya, ''Input Impedance and Radiation Characteristics of Small Microstrip Phased Arrays Including Mutual Coupling for MSAT Applications,'' *Electromagnetics*, Vol. 6, No. 4, 1988, pp. 333–349.

[24] Munson, R. E., ''Conformal Microstrip Antennas and Microstrip Phased Arrays,'' *IEEE Trans. on Antennas and Propagation*, Vol. 22, No. 1, Jan. 1974, pp. 74–78.

[25] Edwards, T. C., *Foundations for Microstrip Circuit Design*, New York, NY: John Wiley & Sons, pp. 86–88.

[26] Ashkenazy, J., S. Shtrikman, and D. Treves, ''Electric Surface Current Model for the Analysis of Microstrip Antennas on Cylindrical Bodies,'' *IEEE Trans. on Antennas and Propagation*, Vol. 33, No. 3, March 1985, pp. 295–300.

Chapter 4
Advanced Feeding Techniques

The previous chapters have dealt with the analysis and design of patch antennas and basic feed approaches. Recently several new feed approaches have been developed that increase the flexibility of microstrip antennas. One feature of these feeds is that there is no direct contact between the feed and the patch antenna. This eliminates the need to use a probe to feed the antenna, which can be an advantage in many cases. Production costs may be reduced because there is no need to fabricate the pin and solder it in place. Reliability is improved as there is no solder joint to break. At high frequencies, pin sizes become small and alignment is very critical. A noncontact feed does not have these problems. This chapter describes electromagnetically coupled patches, aperture coupled patches, and coplanar waveguide fed patches. There is also a very brief discussion of printed dipole and slot antennas.

4.1 LISTING OF COMPUTER PROGRAMS

Specific programs are not given for the design of the antennas presented in this chapter because their analysis is more complicated and beyond the scope of this text. Auxiliary programs that aid in their design are discussed. Several of these programs were initially discussed in previous chapters. The following is a recapitulation of new programs first mentioned in this chapter.

COVMIC: Calculates the characteristic impedance, effective dielectric constant, and attenuation of a microstrip line with a dielectric cover layer. Uses the same quasi-static analysis discussed in Chapter 3 in conjunction with the dielectric covered patch antenna. Used for determining the line parameters of the feed of an emc patch. Inputs are the thickness, effective dielectric constant, loss tangent of both layers, the conductor width, thickness, relative conductivity, and finally the frequency.

CPW2: Determines the characteristic impedance and effective dielectric constant for a coplanar waveguide transmission line. Can analyze a basic coplanar waveguide configuration on a substrate, one with a ground plane on the bottom of the substrate, and one with a ground plane suspended above the substrate, or with both ground planes present. A conformal mapping approach provides a quasi-static solution for the transmission line. Used when designing coplanar waveguide-fed patches. Inputs are substrate thickness and relative dielectric constant, conductor width, and slot width (assumed the same on both sides of the conductor). Unlike MICRO or COVMIC, CPW2 can analyze one width or a range of widths simultaneously.

CPS: Calculates the characteristic impedance, effective dielectric constant, and attenuation for coplanar strips transmission line. Uses the results of a quasi-static analysis with a conformal mapping approach. Used for designing part of the feed for a printed circuit dipole. Inputs are strip width and thickness, substrate height, relative dielectric constant and loss tangent, conductor conductivity, frequency, and spacing between opposing strip edges. Does one or a range of spacings.

SLOTLINE: Finds characteristic impedance and relative dielectric constant for slotline transmission line. Uses closed-form expressions fitted to rigorous full-wave numerical solutions. Used for designing rectangular printed circuit slot antennas. Inputs are substrate thickness and relative dielectric constant, frequency, and slot width. Does one or a range of slot widths.

4.2 ELECTROMAGNETICALLY COUPLED PATCHES

When using a microstrip line as a feed, the line is normally on the same surface as the patch and is directly connected to the patch at one of its edges. The line does not have to directly contact the patch. If the line is on the same surface, it can be brought into close proximity of the patch with a small gap left between the end of the line and the antenna. The capacitance of the gap provides the coupling mechanism to the antenna. Generally the coupling is fairly low, so this approach usually does not offer any advantage over a direct feed.

If the feed is at a lower height than the patch, it can be brought underneath the antenna. The resulting coupling can be very strong. Figure 4.1 shows this feed approach. Two substrates are used. The patch is etched on the top one. The ground plane for this substrate is completely removed. The feed line is etched on the lower substrate, which has its ground plane left intact. The two substrates may have different thicknesses and relative dielectric constants. The feed line is run underneath the patch and left open circuited. The distance, s, from the patch edge to the line edge is referred to as the overlap. The fringing fields from the open-circuited end provide the main coupling mechanism to the antenna. The patch is considered to be *electromagnetically coupled* (emc) to the feed and is often called an emc patch.

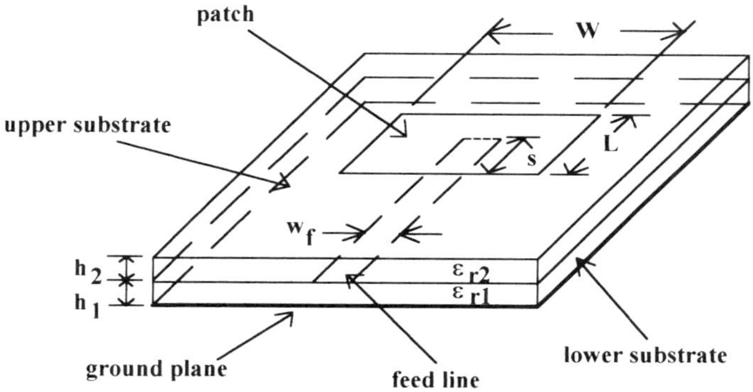

Figure 4.1 An electromagnetically coupled patch antenna.

The coupling mechanism is predominately capacitive. An approximate equivalent circuit at a point on the feed line right at the patch edge is given in Figure 4.2 [1]. The parallel RLC circuit represents the patch itself. C_c is the coupling from the feed to the patch. The coupling is controlled primarily by two factors, the inset distance of the feed and the patch width. The coupling increases with feed inset reaching a maximum when $s = L/2$. The coupling is symmetrical with respect to the center of the patch. Decreasing the patch width increases the coupling. The substrate parameters also have some effect. If the feed substrate has a higher relative dielectric constant, its thickness must be increased to maintain a level of coupling. The effect of substrate parameters is less than that due to the inset and patch width.

With the actual feed point underneath the patch, it is difficult to measure the input impedance at that point. When experiments are performed, the feed line is brought out to some convenient location outside of the patch. Figure 4.3 shows some idealized impedance versus frequency plots that might be measured for an emc patch. Several plots are presented. These would be typical results for a half-wavelength-long feed line. For longer lengths, the loops would rotate in a clockwise direction and the ‘‘tails’’ on either side would

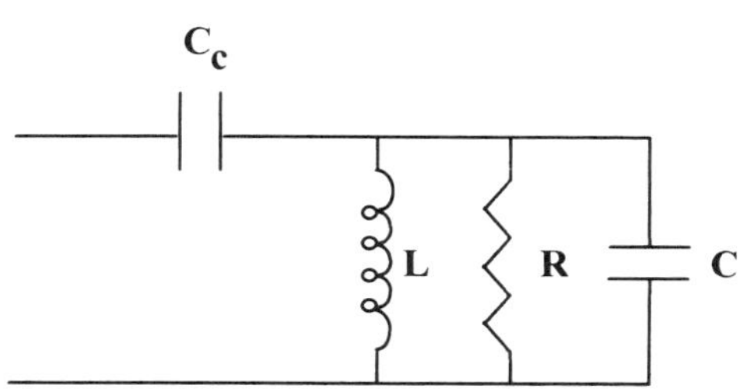

Figure 4.2 Emc patch equivalent circuit at the patch edge.

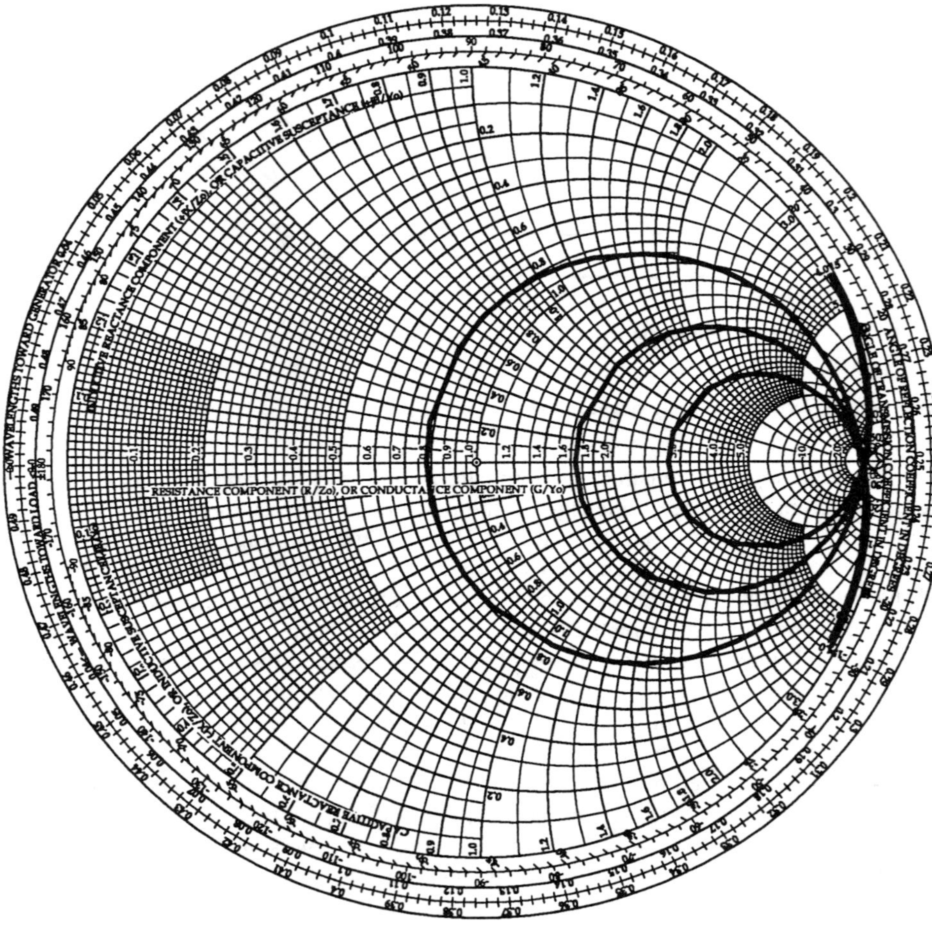

Figure 4.3 Idealized impedance versus frequency plots for an emc patch.

become longer. Note that impedance behaves much like a coupled resonant circuit. The smaller impedance loop at the right represents an under-coupled situation. Conversely, loops to the left of the chart center are over-coupled. The increasing loop sizes represent either increasing feed overlap or decreasing patch width. With proper adjustment of inset and patch width, a matched condition can be obtained.

Generally speaking, if different dielectric constant substrates are used, the lower one has the higher value. This reduces the size of the feed network. Normally using a high dielectric constant substrate degrades patch performance. With two substrates, the overall effective dielectric seen by the patch will be lower because of the upper substrate. Patch performance is thus improved over using a single substrate with the higher dielectric constant. Surface-wave excitation is reduced, and factors such as efficiency and bandwidth are increased. Using different dielectric constant substrates offers a way of meeting the often conflicting requirements of minimizing circuit size while maintaining good antenna performance.

The modeling and analysis of emc patches is quite difficult because of the complicated electromagnetic interaction between the feed and the antenna. Most analyses have been full-wave solutions that are not amenable to design purposes. Recently a modified transmission line model was proposed [2], but even this one involves a fair amount of computational effort to determine capacitances associated with various parts of the circuit. Fortunately the results of a full-wave study have been condensed into a set of design guidelines [3]. These at least provide a good starting point and reduce the amount of experimentation needed. Some experimental results are presented in [4] and [5].

To begin the design, select the substrate thicknesses and dielectric constants. In [3] it is shown that in most cases the lower thickness should be equal to or perhaps slightly less ($\geq 80\%$) than the upper one. To minimize surface-wave excitation, keep the overall thickness of both substrates to less than about $0.04\lambda_0$. Choose a low dielectric constant material for the top substrate unless a very small patch size is required and the decrease in electrical performance can be tolerated. Choose as low a dielectric constant material as is consistent with the area available for the feed circuit for the lower substrate. Usually it is preferable to use the same material for both substrates. Calculate the patch performance using PATCHD. When two different substrates are used, the dielectric constant input needed by the program can be approximated using an average value. With the nomenclature of Figure 4.1, $\epsilon_r \approx [h_1\sqrt{\epsilon_{r1}} + h_2\sqrt{\epsilon_{r2}}]/(h_1 + h_2)$. For a 50% overlap of the feed, the resonant frequency shifts upward by 1% to 2%, so it might be beneficial to design the patch at a correspondingly lower frequency. From the calculations determine if the antenna performance (for example, bandwidth and efficiency) is adequate. In most cases select a patch width that is less than or equal to the length. A width of about 0.5 to 1 times the length is a good starting point. This provides good coupling with the feed. If the antenna performance is inadequate, change either the patch dimensions or the substrate parameters. After doing an initial sizing with PATCHD, consider running program TLPATCH (discussed in Chapter 6) to obtain a better estimate of patch dimensions if different substrate materials are used. When the same materials are used, use PATCH9. In either case, the calculated impedance won't be accurate.

Once an initial set of substrate and antenna dimensions are found, determine the feed line width. Since the feed is a covered microstrip (outside the patch), use COVMIC to find the feed line width for 50Ω and the effective dielectric constant. COVMIC calculates the characteristic impedance and effective dielectric constant for a microstrip line covered with a dielectric. It uses the quasi-static analysis given by (3.19) to (3.22) to find these quantities. Its usage is identical to MICRO except for the need to input the cover layer thickness, relative dielectric constant, and loss tangent. The following is an example of its use.

COVMIC.V40   01-02-1996   22:04:23
[Current Program Date - 25/10/95]

Calculates Zo, eff, and attenuation for a covered microstrip

INPUT SUBSTRATE HEIGHT (cm)? **.16**

(The term substrate here refers to the lower substrate, that is, the substrate underneath the microstrip line.)

INPUT SUBSTRATE RELATIVE er? **2.2**

INPUT SUBSTRATE LOSS TANGENT? **.001**

INPUT COVER LAYER THICKNESS (cm)? **.080**

(This is the dielectric on top of the microstrip line.)

INPUT COVER LAYER RELATIVE er? **4.0**

INPUT COVER LAYER LOSS TANGENT? **.002**

INPUT LINE WIDTH (cm)? **.254**

INPUT CONDUCTOR THICKNESS (cm)? **.00356**

INPUT CONDUCTOR CONDUCTIVITY RELATIVE TO COPPER? **1**

INPUT FREQUENCY (GHz)? **1.575**

(After this last input, COVMIC computes the line characteristics, which are then displayed on the screen.)

Zo = 68.52 (Ohms)
eff = 2.25
attenuation = 0.00551 (dB/cm)

CHANGE LINE WIDTH (w), DO ANOTHER CASE (c) , OR QUIT (q)? **q**

(If **w** is selected, all inputs remain the same. The user enters a new value for the line width. If **c** is chosen, all the inputs are requested again.)

After the feed line width is determined, assume a 50% overlap, that is, $s = L/2$. Make the feed line a half-wavelength long so that it extends out beyond the patch edge.

Build the antenna and measure the impedance as a function of frequency. Be sure to avoid air gaps between the substrates as this introduces repeatability problems. Adjust the patch length to obtain the desired operating frequency. Adjust the patch width to match the antenna. If the frequency is low enough, it may be possible to trim the patch dimensions using a knife or scalpel. Another technique is to use tape to mask off the patch and then to apply a swab dipped in ferric chloride to the area where the copper is to be removed. The trimming technique is best when the minimum amount that can be removed is small enough to give adequate (and repeatable) control of electrical performance. If the trimming approach is not suitable, it may be necessary to build patches with several lengths and widths to fine tune the dimensions. In some cases, some trimming of the overlap may be useful.

One issue with the emc patch is how to treat the upper layer in areas outside the vicinity of the patch. From the patch performance standpoint, the upper layer should be extended over the same area as the lower layer. In this way the patch ''sees'' a uniform environment, which is especially important for the radiation patterns. If extending the upper layer completely over the lower layer is not feasible, then make it as large as possible. When the circuits connected to the patch are on the lower layer, the effect of the upper layer should be included in calculating line lengths and other dimensions. The presence of the upper layer is, of course, awkward if components such as resistors and capacitors are present. Obviously, the upper substrate must be truncated. Again, truncate it as far away from the patch as feasible and at a point where there is minimum circuitry such as only a single-feed line. Use as low a dielectric constant as possible for the upper layer. In this way there is the smallest change in the electrical environment. Any changes in line width, for example, will be minimal.

In [1] it was found that the emc patch bandwidth can be increased significantly by proper placement of a short shunt mounted, open-circuited stub on the feed line. An SWR = 2.0:1 bandwidth of up to 13% has been obtained. The relevant geometry is shown in Figure 4.4. The stub is located a little over a half-wavelength (as measured on the feed line) down the line from the patch edge. The stub length is about 0.10 to 0.12 a microstrip wavelength long. The stub width is not specified but appears to be the same as that of the feed line. The purpose of this stub is to position the impedance locus (such as seen in Figure 4.3) more in the center of the Smith chart. This increases the impedance bandwidth. The stub introduces a slight ripple in the *E*-plane pattern at some frequencies. Keep the stub length as short as possible to minimize both this effect and any cross polarization. The actual dimensions should be found by experiment.

A similar bandwidth widening can also be obtained with probe-fed emc patches. The configuration is shown in Figure 4.5. An appropriate probe-to-microstrip transition,

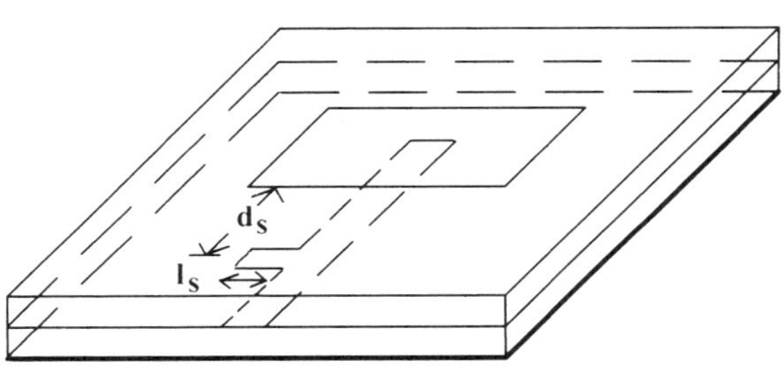

Figure 4.4 Emc patch with matching stub for broad bandwidth performance.

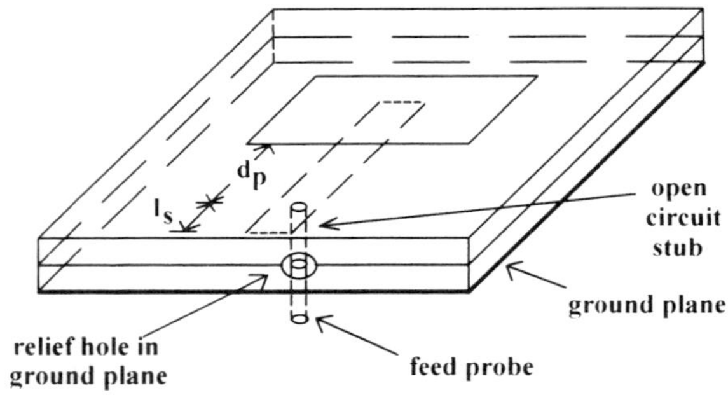

Figure 4.5 Emc patch with probe feed and stub.

as discussed in Section 5.9 (see Figures 5.15(b) and 5.16), is placed again about just over a half-wavelength away from the patch edge. The open-circuited stub, which is part of the transition, not only ensures proper transition operation but also serves as a tuning element to broaden the antenna bandwidth. A good starting point for the stub length is about a quarter-wavelength. Both the probe position, d_p and stub length, l_s are determined experimentally.

4.3 APERTURE COUPLED PATCHES

Another type of noncontacting feed for microstrip patches uses a rectangular aperture in the ground plane as the coupling mechanism. This was first proposed by Pozar [6]. A sketch of an aperture coupled patch is given in Figure 4.6. The patch is etched as normal on a substrate. On the ground plane underneath the patch, a rectangular slot is etched. The slot is placed at the center of the patch. The feed line is on another substrate where the ground plane is completely removed. The slot is considerably smaller than the patch. The two substrates are placed on top of each other aligned so that the feed line crosses over the center of the slot. The feed extends beyond the slot usually about a quarter-wavelength. As indicated in Figure 4.6, the substrates can have different heights and dielectric constants.

Aperture coupling offers several advantages. One is that the patch is completely shielded from the feed. Spurious radiation from the feed does not degrade side lobe levels or increase cross polarization. The individual substrates can be tailored for optimum performance. The patch requires a low dielectric constant for larger bandwidth and less surface-wave excitation. A high dielectric constant for the feed substrate reduces the circuit size. Because of the intervening ground plane, the dielectric constant of one does not affect the other. Like the emc patch, there are several additional parameters at the designer's disposal. The aperture length and feed line extension are the two major parameters.

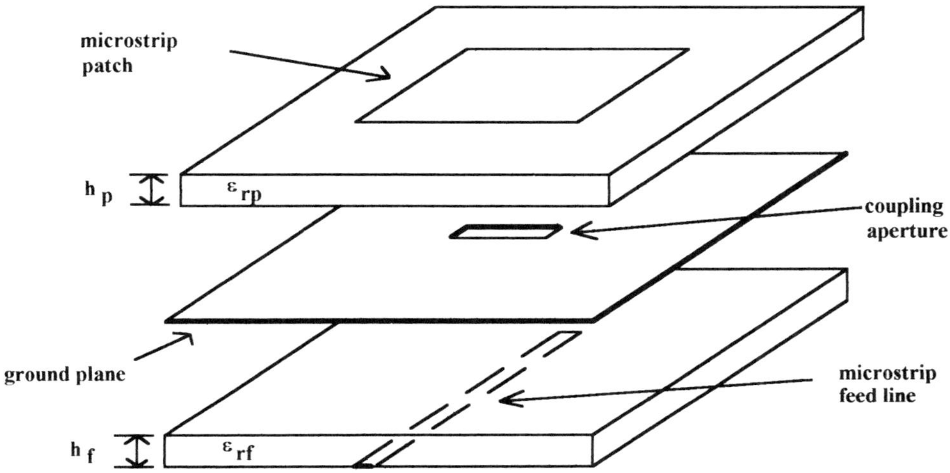

Figure 4.6 Aperture coupled microstrip patch.

A more detailed sketch of the aperture and feed line is presented in Figure 4.7. The slot is shown off-center even though it is normally centered. The feed line is along the resonant dimension of the patch. The slot, therefore, has its length perpendicular to the resonant dimension. The feed line is narrower than the slot.

When fed with aperture coupling, the patch is in series with the feed line. The aperture is too small to be resonant, so it contributes only a reactance to the impedance. The stub beyond the slot also presents a series reactance. An equivalent circuit for the antenna is shown in Figure 4.8 [7]. L' is the inductance associated with the below resonance slot. The patch is the parallel RLC circuit. The stub, of course, is an open circuited transmission line with the same characteristic impedance as the feed line. The stub compensates for the inductance of the slot and the patch to help create a real input impedance for the antenna.

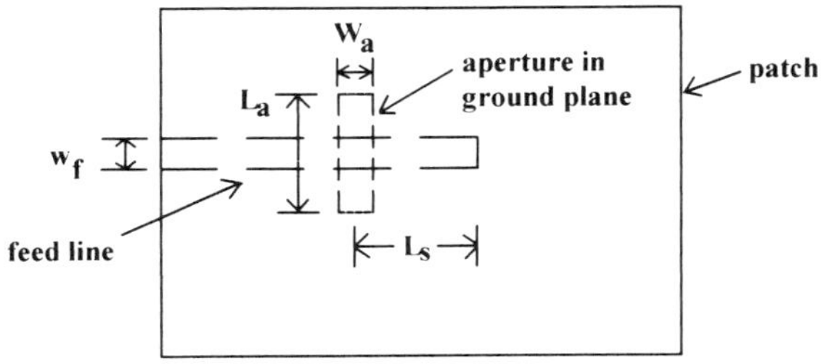

Figure 4.7 Relationship between patch, aperture, and feed line.

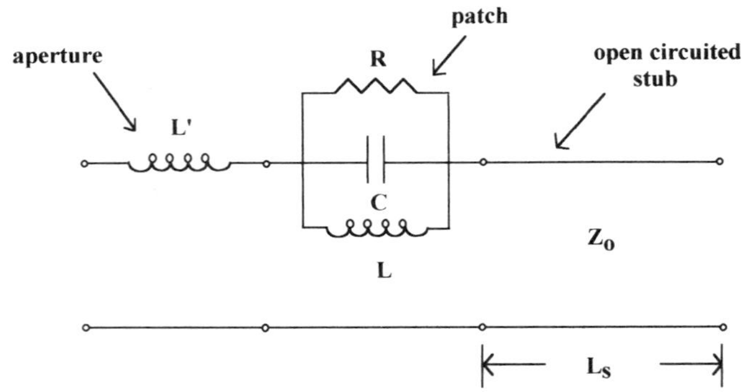

Figure 4.8 Equivalent circuit for an aperture coupled patch.

The behavior of the input impedance with changes in stub length is shown in Figure 4.9. As the stub length increases, the impedance locus rotates in a clockwise direction along a constant resistance circle [8]. As the stub length approaches a quarter-wavelength long, the impedance curve crosses the real axis of the Smith chart. The stub length is adjusted until the impedance at the design frequency is purely real. This is illustrated by the leftmost curve in Figure 4.9. With the correct stub length, the impedance is not only real at the design frequency but also symmetrical about the real impedance line.

While the stub length rotates the impedance locus, the aperture length controls the amount of coupling. Typical impedance curves as a function of aperture length are presented in Figure 4.10. As the aperture length increases, the curve moves toward the right [8]. By adjusting the length, the impedance at the design frequency can be set to 50Ω or some other convenient value. A similar behavior occurs as the slot is moved in the direction of resonance of the patch [8]. As the slot moves away from the center, the impedance locus shifts to the left. Usually the slot is left in the center, and the aperture length is used to adjust the coupling. Surprisingly, the impedance changes very little for slot movements in the direction perpendicular to the patch resonance as long as the slot remains completely underneath the patch [8].

The feed substrate thickness and dielectric constant also affect the coupling. As the substrate gets thicker, less coupling occurs [8]. A longer slot is needed to maintain a given level of coupling. The impedance moves to the left as in Figure 4.10. The opposite happens as the dielectric constant increases [8]. The required slot size therefore decreases. As the slot size increases, so does its radiation. This not only causes power to radiate on the wrong side of the ground plane but also reduces patch efficiency. Similar coupling behavior occurs with changes in the antenna substrate thickness and dielectric constant.

As with emc patches, the analysis of aperture coupled patches is very complicated. A very complete full-wave analysis has been published [8]. Cavity-based models have

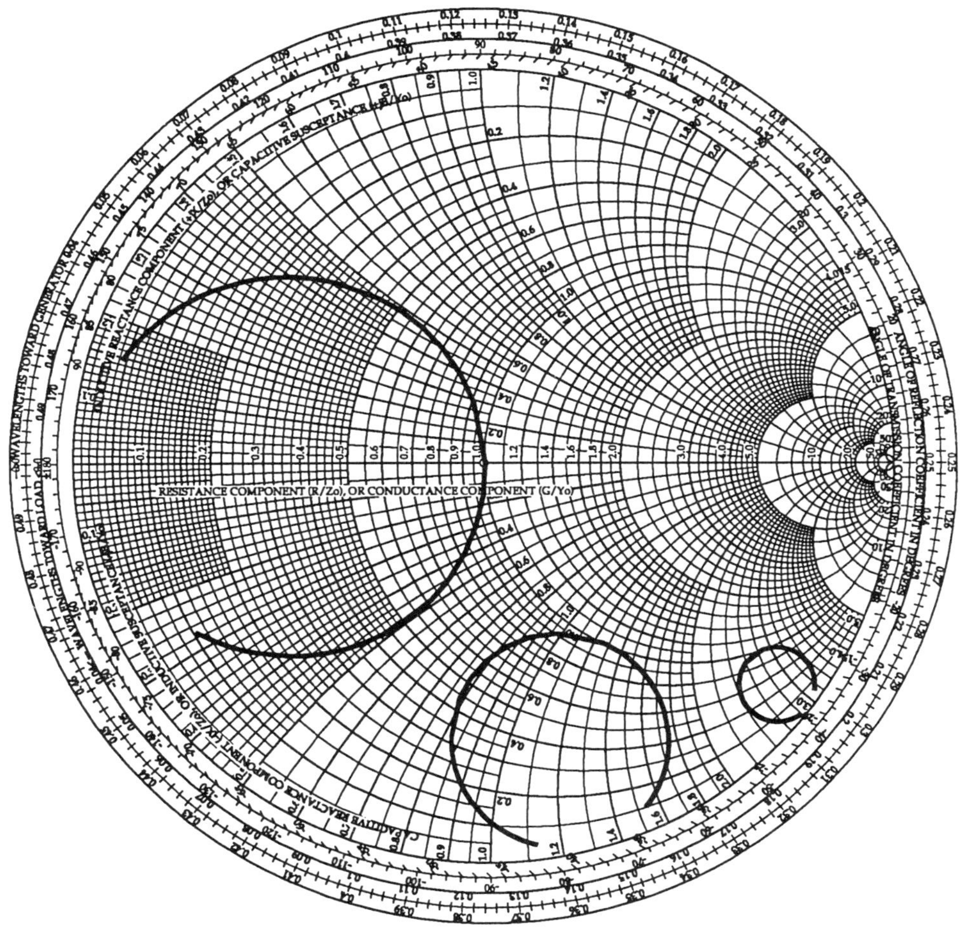

Figure 4.9 Aperture coupled patch impedance as the stub length is varied.

also been developed [9,10]. Even these require a fair amount of computational effort. Both [8] and [9] provide design information and guidelines that will prove useful. To begin the design, select the substrates. Unless a minimum patch size is required, choose a low dielectric material for the patch substrate. The feed substrate can be selected to realize the needed feed circuit size. Keep the substrate thicknesses small, on the order of $0.01\lambda_0$ to $0.03\lambda_0$ for each [8]. This keeps the slot size from getting too large and causing too much radiation.

Based on the results in [8] and [9], the patch width should be chosen to be about 0.75 to about 0.875 times the patch length. Use PATCH9 to design the antenna. The

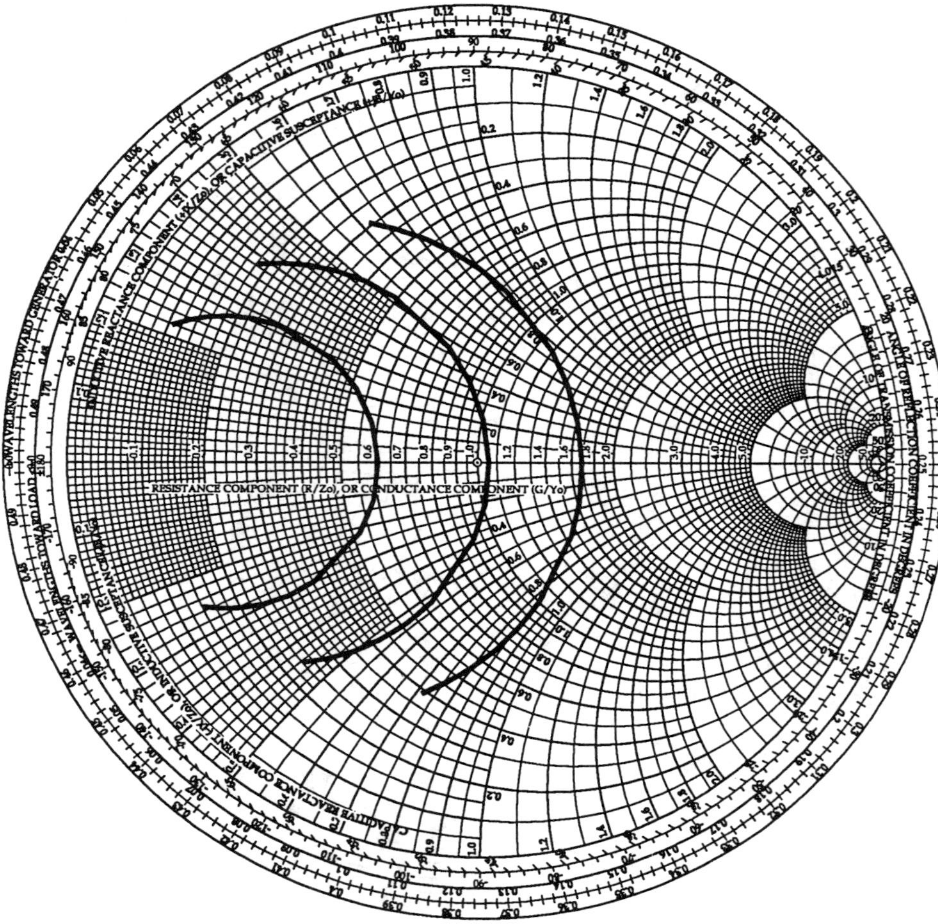

Figure 4.10 Changes in impedance with different aperture lengths.

aperture coupling reduces the resonant frequency between 2% to 4%. Design the patch to operate at an appropriately higher frequency. Position the coupling aperture in the center of the patch. For antenna and feed substrates that are each on the order of $0.01\lambda_0$, the slot length is about $0.082\lambda_0$ for low dielectric constant substrates ($\epsilon_r \approx 2.54$). If a higher dielectric constant feed substrate is used, the slot length decreases to $0.074\lambda_0$ for an $\epsilon_r \approx 10.2$. Should the patch substrate have a higher dielectric constant (than the feed), the slot size would decrease even more. The slot width is kept small for best coupling. Choose one that is about a tenth of the length. For thicker substrates, the slot length must be increased from these estimates. The stub length beyond the center of the slot is around $0.22\lambda_f$ where λ_f is the wavelength on the feed line. Use MICRO to determine the feed line width (usually chosen to produce a 50-Ω line) and the line effective dielectric constant.

The final dimensions need to be determined experimentally. Keep the patch size, feed line width, and slot width constant. Build several antennas with differing slot and stub lengths. Results from measurements on these antennas can be used to ascertain the final dimensions. In the case of the stub length, the antennas may be purposely built with longer stubs. They can be trimmed back to determine the optimum length for a given slot length.

4.4 COPLANAR WAVEGUIDE-FED PATCHES

Although microstrip and stripline are, by far, the most widely used printed microwave transmission media, others have been developed over the years. One that is now receiving some attention is coplanar waveguide [11]. This is shown in Figure 4.11. The entire transmission line is on one side of the substrate. Two narrow strips of metalization are etched away to leave a center conductor of width *w*. The edges of the center conductor are separated from the remaining metalization by a gaps of width *s*. The sections of metal on either side are ground. The electric field lines loop across the gap from charges on the center conductor to charges of opposite sign on the ground metalization. Magnetic field lines form ellipses that encircle the center conductor.

One of the reasons for the current interest in coplanar waveguide is that it is easy to incorporate both series and shunt elements [12]. The former is inserted by breaking the center conductor and mounting the component across the resulting gap. Shunt elements are mounted across the gap between center conductor and ground. This capability also applies to transmission line structures, so it is possible to realize both distributed filter circuits as well as series and shunt reactances. Coplanar waveguide is thus ideal for on-

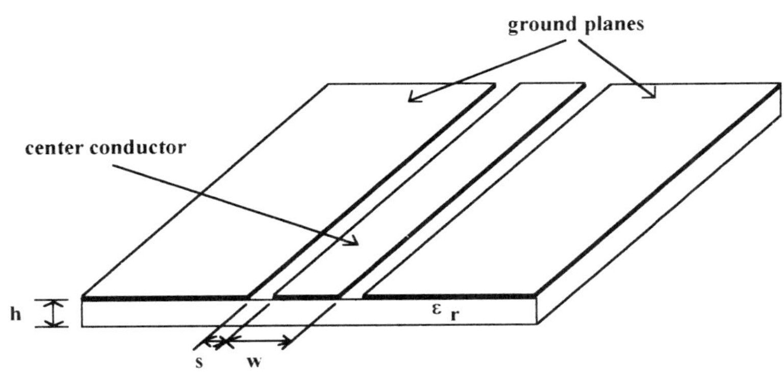

Figure 4.11 Geometry for coplanar waveguide.

chip circuits where shunt mounted elements are difficult to realize with microstrip. One does not need to use vias to ground shunt elements. Having the entire transmission line on one surface does have one major drawback. The fields are not "shielded" on one side by a ground plane. The line characteristics can thus be influenced by the presence of package lids and bottoms. If the substrate is laid on the bottom of a package, like a microstrip circuit would be, the line impedances, for example, will change. Either the substrate should be kept away from large metal surfaces or the presence of the other surface accounted for in the design.

The impedance and effective dielectric constant are a function of the center conductor width, gap spacing, substrate dielectric constant, and substrate thickness. For a fixed center conductor width, the impedance increases with increasing gap width and vice versa. The impedance decreases with increasing dielectric constant. It also decreases somewhat with increasing substrate thickness. Program CPW2 calculates the characteristic impedance and effective dielectric constant for coplanar waveguide. It uses a quasi-static approach based upon a conformal mapping method [13]. Program use is very straightforward.

PROGRAM CPW2.V10 11/15/95 08:35:13
[Current Program Date - 11/15/95]

ANALYZE
(1) - CPW no ground plane
(2) - CPW with lower ground plane
(3) - CPW with upper ground plane
(4) - CPW with upper and lower ground plane
1

(As this indicates, the program can account for additional ground planes both above and below the substrate. The lower ground plane is assumed on the bottom side of the substrate. The upper one is separated from the coplanar waveguide surface by an air filled layer. If options (3) or (4) are selected, the program asks for the height of the upper ground plane above the top surface of the substrate.)

INPUT DIELECTRIC SUBSTRATE THICKNESS (cm)? **.16**

INPUT DIELECTRIC SUBSTRATE RELATIVE DIELECTRIC CONSTANT? **2.2**

INPUT CENTER CONDUCTOR WIDTH (cm)? **.0635**

INPUT NUMBER OF SLOT WIDTHS TO ANALYZE? **10**

(If this is answered with the number **1**, then the program requests that width in centimeters. If more than one width is requested, the user is given the option of entering range of widths or each width individually. The former will be illustrated here.)

INPUT RANGE OF WIDTHS (r) OR INDIVIDUAL WIDTHS (I)? **r**

INPUT SMALLEST AND LARGEST SLOT WIDTHS (cm)? **.00254,.0279**

The impedance and effective dielectric constant are calculated and displayed on the screen as follows.

PROGRAM CPW2.V10 10-15-1995 20:04:32
SUBSTRATE THICKNESS = 0.160 (cm)
RELATIVE DIELECTRIC CONSTANT = 2.20
CENTER CONDUCTOR THICKNESS = 0.0635 (cm)

COPLANAR WAVEGUIDE

SLOT WIDTH (cm)	Zo (ohms)	eff
0.00254	50.47	1.60
0.00508	58.80	1.59
0.00762	64.85	1.59
0.01016	69.78	1.59
0.01270	74.01	1.59
0.01524	77.76	1.59
0.01778	81.15	1.59
0.02032	84.25	1.59
0.02286	87.12	1.59
0.02540	89.79	1.59
0.02794	92.30	1.59

The final input is

PRINT OUT RESULTS (p), STORE RESULTS IN A FILE (s), DO ANOTHER CASE (a), OR QUIT (q)? **s**

When answered with an **s**, the next input is the name of the file in which the results are stored.

A coplanar line can be used to excite a patch antenna [14]. The coplanar line is placed in the ground plane of the patch substrate. Two examples are given in Figure 4.12. In one the coupling is inductive; the center conductor is connected directly across the slot. In the other, the center conductor is stopped at the slot to form a capacitive coupling.

Both can be used with similar results. The slot length for the inductive coupling is somewhat more critical. Experimental results [14] show that with a substrate having an $\epsilon_r = 2.22$, the optimum slot length is about $0.245\lambda_s$ where λ_s is the wavelength in the slot. This is found by approximating the effective dielectric constant in the slot by $\epsilon_{eff} \approx (1 + \epsilon_r)/2$. The coupling slot introduces some inductance and therefore reduces the patch resonant frequency. For the inductive coupling, the reduction is about 3.2%. With the capacitive coupling, the slot length is less critical, ranging from $0.250\lambda_s$ to $0.288\lambda_s$. The corresponding resonant frequency decrease is from 14% to 24%. Obviously, the capacitive slot reduces the frequency more than the inductive one. The bandwidth for an SWR = 2.0:1 is about 3.5% for the inductive coupling and 2.8% for the other.

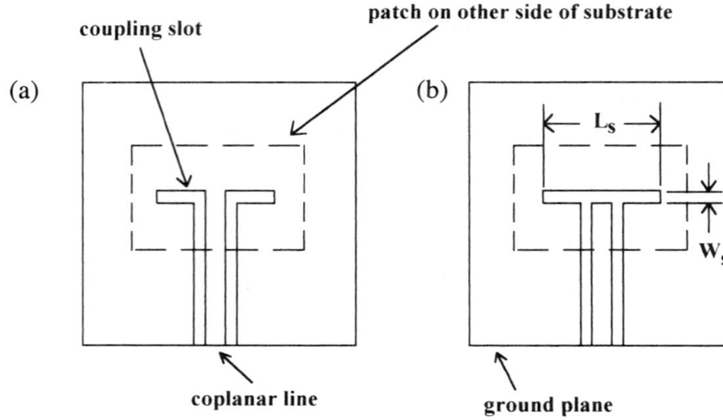

Figure 4.12 Coplanar waveguide fed microstrip antennas: (a) inductive coupling and (b) capacitive coupling.

When designing this antenna, design the patch as before using PATCH9, except use a higher frequency to compensate for the slot. Use CPW2 to find the dimensions of a 50-Ω coplanar line. Start with the above estimates for the slot length, make the slot width about one-tenth the width. Experimentally optimize the slot dimensions. One advantage of a coplanar feed is that it is relatively easy to interface with a coaxial connector as seen in Figure 4.13. A flange mount connector can be butted directly up to the substrate. Just make sure that the center conductor width is less than the diameter of the dielectric in the connector to prevent shorting out the coplanar line.

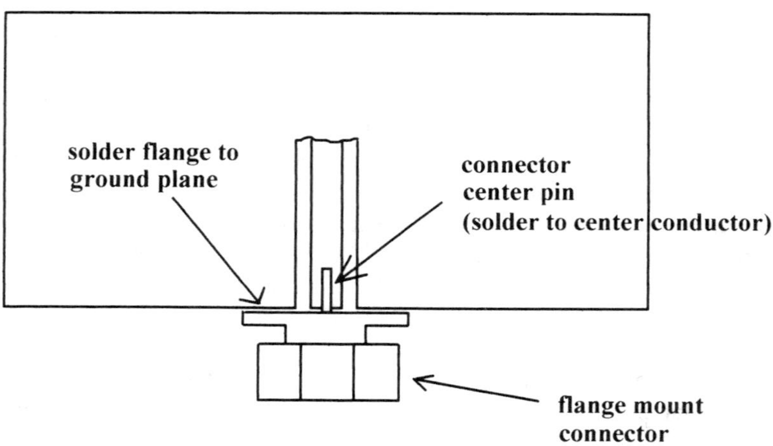

Figure 4.13 Coplanar waveguide to coaxial connector transition.

One disadvantage of the above coplanar feeds is radiation from the slot results in a relatively poor front-to-back ratio. The pattern level on the ground plane side is only 9 dB to 10 dB down from the peak of the patch pattern [14]. This is because the slots are electrically quite long and so are relatively good radiators in their own right. If the slot is made into the shape of a loop, then its size can be reduced, which eliminates much of the back radiation [15]. Figure 4.14 shows the geometry. For the results presented in [15], the substrate dielectric constant is 10.8, so an approximate slot ϵ_r is about 5.9. The loop diameter, d, is about $0.095\lambda_s$ where again, λ_s is the slot wavelength. No explanation is given as to why this diameter was chosen. The loop is centered underneath the patch.

4.5 OTHER TYPES OF PRINTED CIRCUIT ANTENNAS

Although the focus of this book is microstrip patch antennas, there are a number of other types of printed circuit antennas. In some applications, these other types may be more appropriate than patches. This section briefly discusses some of these antennas.

There are several variations of printed dipole antennas. Figure 4.15 shows one that can be placed at the end of a circuit board [16]. The dipole is etched on the ground plane side of the board. The ground plane is truncated as shown except for the section of length h_d, which spaces the dipole away. A slit is placed in the middle of this section. The slit serves two purposes. One is to separate the two arms of the dipole. The other is to form a coplanar transmission line comprised of the two sides of the section. This type of transmission line is called coplanar strips. It is very similar to the twin lead often used to connect an antenna to a television. The feed line is routed up the center of one of the strips until it reaches the dipole. There it is bent at a right angle, crosses over the gap between arms, and then bends back down ending in an open circuit.

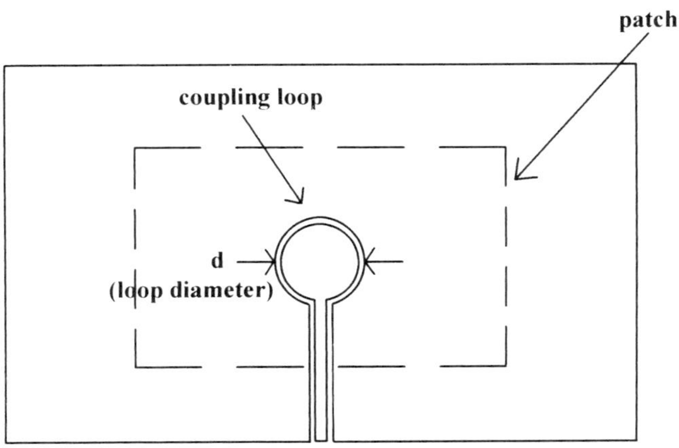

Figure 4.14 Circular coupling slot for reduced feed radiation.

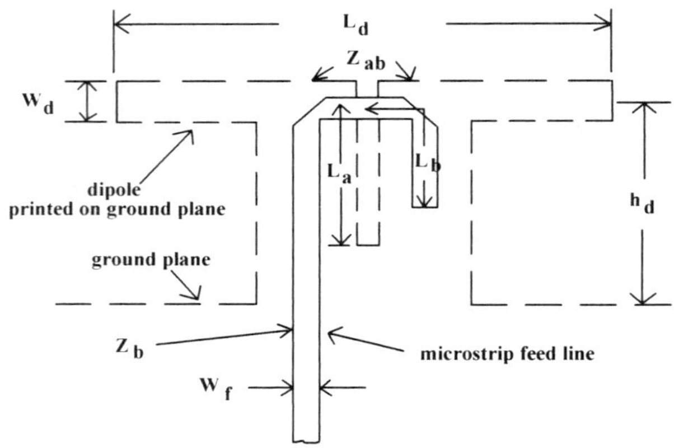

Figure 4.15 Printed circuit dipole with balun.

The reason for this elaborate feed structure is that a dipole is a balanced structure and requires a balun if fed with an unbalanced transmission line, in this case, the microstrip. The balun is formed by the coplanar strips and the open stub after the dipole feed point. An equivalent circuit for this structure can be seen in Figure 4.16.

The microstrip stub is in series with the feed line. The coplanar strips are short circuited and in shunt across the dipole. The coplanar line is approximately a quarter-wavelength long. Being short-circuited, it places a very high impedance at the dipole feed point. In addition, the stub is about a quarter-wavelength long and thus looks like a short circuit. Both of these circuits prevent currents from the microstrip line from flowing back along the ground plane. The only place for them to go is along the dipole arms. This produces the balun action. They also provide a double tuning action that helps match the structure.

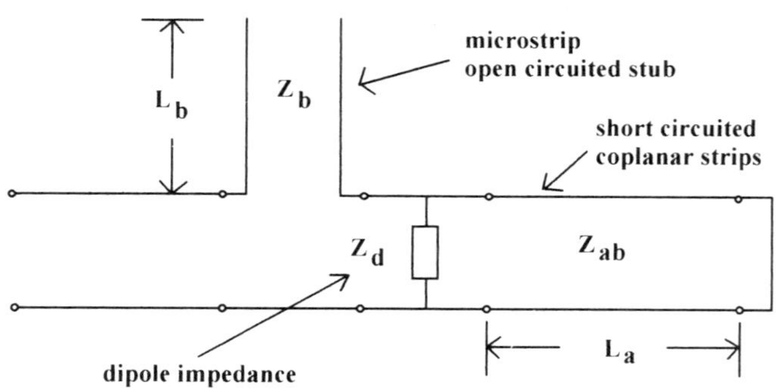

Figure 4.16 Equivalent circuit for dipole and balun.

For relatively low substrate dielectric constants, the dipole impedance can be approximated by that of a free-space dipole. Standard formulas for wire dipoles as contained in most antenna texts [17] may be used for this purpose. To evaluate the equations, the dipole radius must be known. A very thin strip has an equivalent radius, $a_e = 0.25W_f$, where W_f is the strip width. Unfortunately the presence of the ground plane alters the impedance in a difficult to calculate manner. In [16], it was found that with a substrate ϵ_r of 3.8, the dipole was resonant when its length of about $0.4\lambda_0$ for a width of $0.05\lambda_0$. The impedance at resonance was around 80Ω, which is actually not too far from a halfwave free-space antenna. The length is shortened somewhat by the presence of the substrate dielectric.

For the microstrip feed lines to work correctly, the width of the coplanar strips should be at least three times the microstrip line width [16]. The strip widths are constrained by the dipole arms. The total width of the strips should be less than about one-third the dipole length. The minimum gap width is limited by etching accuracies. Conversely, the microstrip line width must be less than a one-third the coplanar strip width. All of these factors restrict the range of realizable impedances to between 60Ω and 100Ω. In [16], both the microstrip and coplanar line impedances were chosen equal to the dipole resonant impedance. Use MICRO to find the microstrip impedance and effective dielectric constant.

The characteristics of the coplanar strips are found using program CPS. CPS operates exactly the same as CPW2 except the dimensions to be input are the strip width and thickness, spacing between strips (as in Figure 4.17), frequency, loss tangent, and conductivity data since attenuation is calculated. The program uses the equations given in [12] to evaluate the characteristic impedance and relative dielectric constant. A sample case for CPS is given as follows.

PROGRAM CPS.V20 01-02-1996 21:56:07
[Current Program Date - 16/10/95]

Calculates impedance of coplanar strips

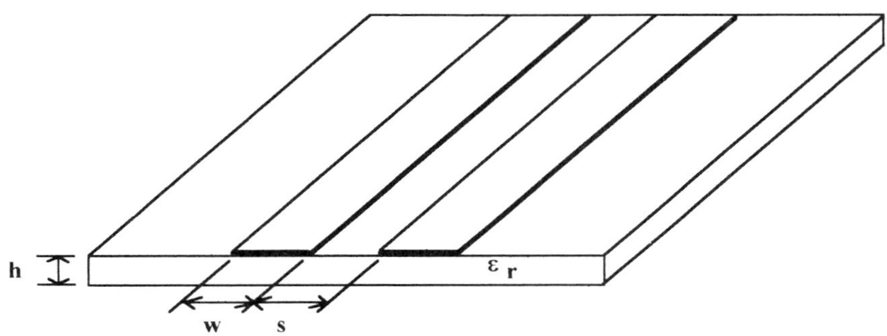

Figure 4.17 Coplanar strips geometry.

INPUT STRIP WIDTH (cm)? **.127**

INPUT CONDUCTOR THICKNESS (cm)? **.00356**

INPUT SUBSTRATE THICKNESS (cm)? **.16**

INPUT RELATIVE DIELECTRIC CONSTANT AND LOSS TANGENT? **2.2,.001**

INPUT CONDUCTIVITY RELATIVE TO COPPER
(Default = 1)
?<**rtn**>

(Hitting the return key results in the default value of 1 for the relative conductivity.)

INPUT FREQUENCY (GHz)? **1.575**

INPUT NUMBER OF SPACINGS? **5**

INPUT RANGE OF SPACINGS (r) , OR INDIVIDUAL SPACINGS (I)? **r**

(When answered with the letter **i**, the program asks for individual spacings, one at a time until all spacings are input. When answered with **r**, the following input is requested.)

INPUT SMALLEST AND LARGEST SPACINGS (cm)? **.0254,.1524**

The computations are performed with the results displayed on the screen as follows.

PROGRAM CPS.V20 01-02-1996 22:03:48

STRIP WIDTH = 0.127 (cm)
CONDUCTOR THICKNESS = 0.00356 (cm)
RELATIVE DIELECTRIC CONSTANT = 2.20
DIELECTRIC THICKNESS = 0.160 (cm)
FREQUENCY = 1.5750 GHz

SPACING (cm)	Zocs (ohms)	er	alpha (dB/cm)
0.0254	892.07	1.515	0.0641
0.0508	891.55	1.517	0.0297
0.0762	893.70	1.509	0.0193
0.1016	896.54	1.500	0.0144
0.1270	899.62	1.489	0.0116
0.1524	902.77	1.479	0.0098

The results can be printed out or stored in a file. The length of the coplanar section, l_a, is such that the electrical length of that section is between 90 degrees to 95 degrees. l_b is chosen to yield an electrical length of 105 degrees to 110 degrees. Since the dipole impedance is not 50Ω, a quarterwave transformer may be added in series with the microstrip line at the balun input. The input impedance of the equivalent circuit of Figure 4.16 is [16]

$$Z_{in} = -jZ_b \cot(k_m L_b) + \frac{jZ_d Z_{ab} \tan(k_{cp} L_a)}{Z_d + jZ_{ab} \tan(k_{cp} L_a)} \tag{4.1}$$

where $k_m = 2\pi\sqrt{\epsilon_{rem}}/\lambda_0$ and $k_p = 2\pi\sqrt{\epsilon_{rep}}/\lambda_0$, with ϵ_{rem} and ϵ_{rep} being the effective dielectric constants of the microstrip and coplanar strip lines, respectively.

This type of printed dipole radiates with a maximum in the plane of the substrate but perpendicular to and away from the ground plane. There is another type of printed dipole with one arm on each side of the substrate [18]. The geometry is shown in Figure 4.18. The dipole is fed with a balanced twin lead transmission line consisting of two strips, one on each side of the substrate. Each strip is bent at a right angle to form the dipole arms.

The conversion from a microstrip line to the twin lead is done with another type of balun as seen in Figure 4.19. The input is a 100-Ω microstrip line. The ground plane is gradually tapered down toward the microstrip line as shown in the figure. In the meantime the microstrip line is widened. The width of the twin lead strips is such that the characteristic impedance is equal to the dipole resistance at resonance. If the substrate thickness is h, it can be shown [18] that the impedance of the twin lead is equal to that of a microstrip line of the same width but with a substrate having half the thickness. The reason why a 100-Ω microstrip line is used is because in [18] the balun was part of an array feed and always at a point where two feed lines joined together. Two baluns from each line were connected together to produce 50Ω. A coaxial connector was then inserted to interface with test equipment. If only one feed line is needed either use a transformer to convert from 50Ω to 100Ω or redesign the balun starting with a 50-Ω line.

Reference [18] offers some empirical suggestions regarding the design of the dipole and balun. The length of the taper section should be at least a wavelength at the lowest operating frequency. The opening angle should be kept as small as possible. In [18], an

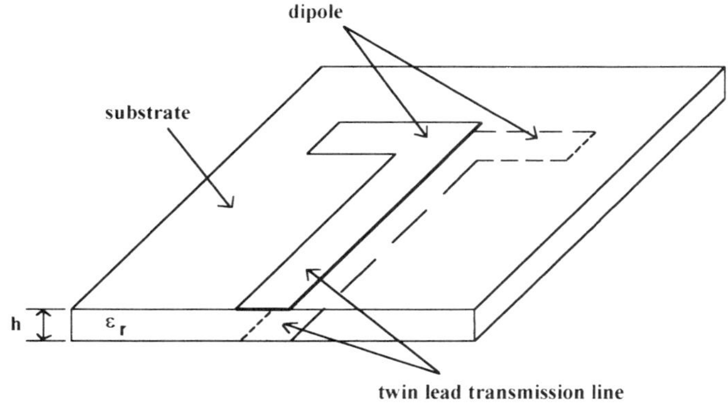

Figure 4.18 Another form of printed circuit dipole.

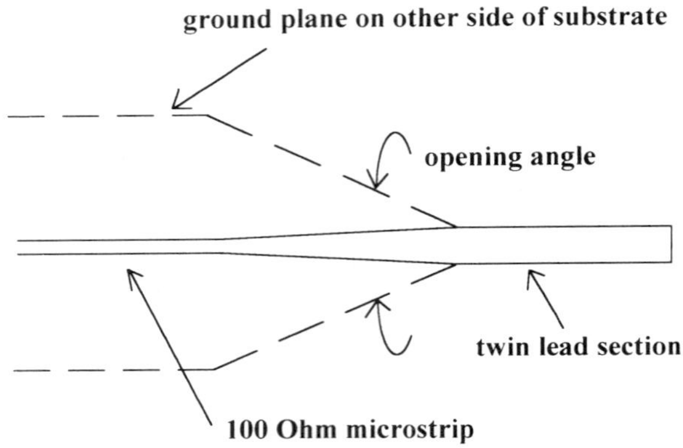

Figure 4.19 Balun for dipole of Figure 4.18.

angle of 6 degrees was found satisfactory. The thickness of the substrate should be kept as thin as possible. This minimizes the gap between the dipole arms. The gap at the dipole input region is the feed point of the antenna. To approximate an ideal voltage source feed more closely, the gap needs to be small. Each dipole arm is about a quarter-wavelength long. Use the effective dielectric constant of the twin lead section to help find the approximate length of the arms. Fairly wide dipoles (width $\approx 0.15\lambda_0$ to $0.20\lambda_0$) were used in [18] to obtain wide bandwidths. If the dipole width is much wider than the feed lines, taper the dipole down toward the feed line width at the input. Taper the dipole over the first quarter or so of the length using a taper angle of 60 degrees or more.

Both types of dipoles may be mounted over a ground plane to produce radiation in one direction. From the antenna of Figure 4.15, the circuit board is protruded through a hole in a metal sheet such that the sheet is perpendicular to the board ground right at the point of truncation. The board ground is connected to the sheet. Make the hole large enough so that the microstrip feed line can pass through without disrupting its fields. For the dipole of Figure 4.18, simply place the substrate parallel to a metal sheet and space it about a quarter-wavelength away.

The baluns of Figures 4.15 and 4.19 can also be used to feed a printed loop antenna. For the balun of Figure 4.15, the dipole arms bent and connected together to form either a circular or square loop. If the other balun is used, a via is needed to connect one end of the loop to one of the twin lead strips. Loops are usually not designed to be resonant. Most loops that are used are electrically small; that is, the circumference is very small compared to a wavelength. The resistance of such a loop is very low, so typically a step-up transformer is employed. Electrically small loops have an inductive reactance that can be tuned out by a capacitor. Calculation of the loop impedance is not very accurate unless quite sophisticated models are used. Generally it is easier to simply experimentally determine the transformer turns ratio and tuning capacitance.

Aperture antennas have been used with microstrip circuits. Figure 4.20 illustrates a rectangular slot fed by a microstrip line. The slot is etched in the microstrip ground plane. A microstrip line is run over the slot to couple a signal into the antenna. The line is extended beyond the slot by approximately a quarter-wavelength and left open circuited. This is done to effectively place a short circuit at the slot edge and maximize the electric field across the slot. Instead of extending the line, a via may be used to short the line right after crossing the slot. The slot length is about a half-wavelength for resonance. The presence of the feed line causes a reduction in the electrical length of the slot. The short circuit at each end of the slot also reduces the electrical length. Typical slot widths are about a tenth or less of the length.

An approximate design technique for microstrip fed rectangular slots is presented in [19]. It is based upon the equivalent circuit of Figure 4.21. The slot is represented by two sections of short-circuited slotline. The conductance, G_s, represents the power lost due to radiation from the antenna. The coupling of the slot to the microstrip line is through the transformer with its self (L_1 and L_2) and mutual (M) inductances. A series-connected transmission line models the open circuited microstrip stub beyond the slot. The end-effect extension is accounted for by the capacitor, C_m, at the end of the line. Any radiation from the open circuit is given by the conductance, G_m. For a given geometry and substrate parameters, the equivalent circuit may be used to calculate the impedance seen by the microstrip feed line. Dimensions are adjusted until the impedance is real and of the desired magnitude at the design frequency.

Begin the slot design by selecting a substrate. The main effects of the substrate are on the slot length, microstrip width, and, to a lesser degree, the radiation patterns. The slot length decreases with increasing dielectric constant. Make sure that the length is significantly greater, say at least four to five times, than the microstrip width. Usually

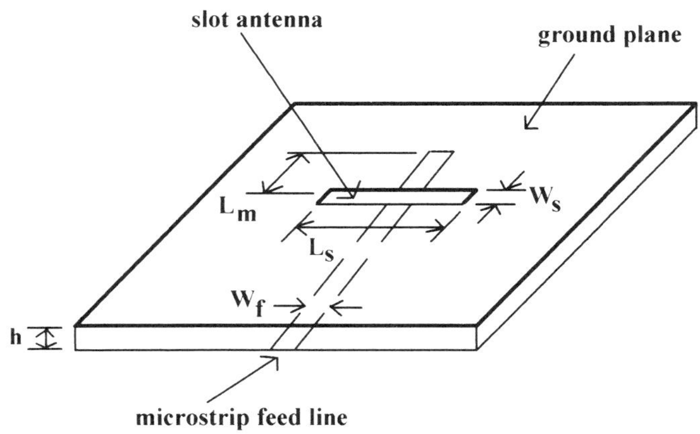

Figure 4.20 Microstrip fed slot antenna.

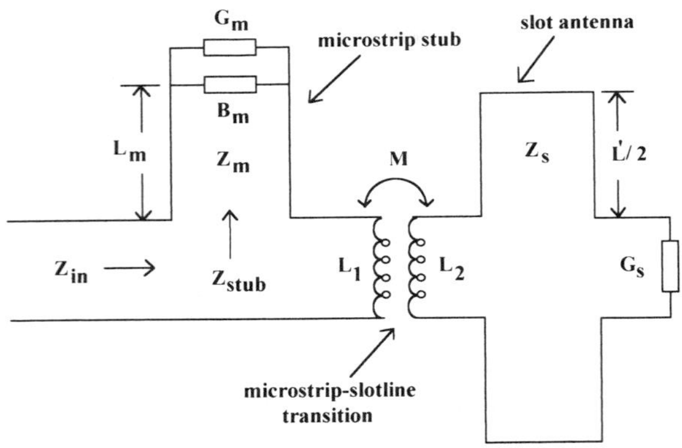

Figure 4.21 Equivalent circuit for a microstrip fed slot.

the microstrip line is taken to have a 50-Ω characteristic impedance. If the slot length becomes comparable to the line width, its behavior may be difficult to predict. If the dielectric constant is very high, say 10 or greater, more power will be radiated on the substrate side of the slot. Some of this power can get trapped in the substrate and travel to other parts of the circuit causing unwanted coupling. As usual, thinner substrates reduce surface-wave problems for both the feed line and slot.

Use program SLOTLINE to calculate the characteristic impedance and effective dielectric constant of the slotline transmission line in the model. A sketch of the slotline geometry is given in Figure 4.22. SLOTLINE uses expressions presented in [12] and [20], which are essentially closed-form curve expressions fitted to more rigorous analyses results. It follows the same format as CPW2. The microstrip parameters are found using MICRO.

A sample case for SLOTLINE is given as follows.

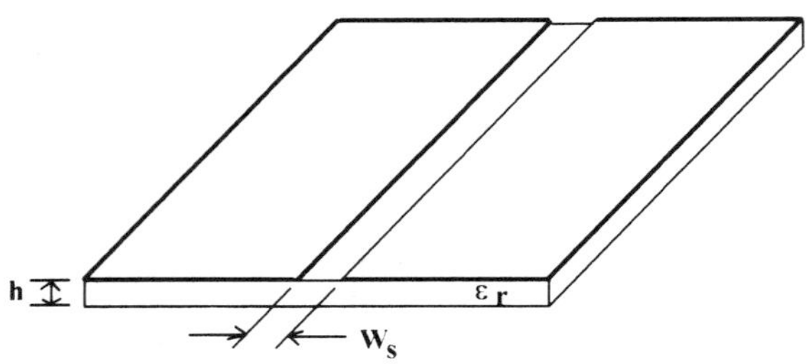

Figure 4.22 Slotline geometry.

PROGRAM SLOTLINE01-02-1996 22:06:28
[Current Program Date - 17/10/95]

Calculates Zo and eff for slotline

INPUT DIELECTRIC SUBSTRATE THICKNESS (cm)? **.16**

INPUT DIELECTRIC SUBSTRATE RELATIVE DIELECTRIC CONSTANT? **2.2**

INPUT FREQUENCY (GHz)? **1.575**

INPUT NUMBER OF SLOT WIDTHS TO ANALYZE? **10**

INPUT RANGE OF WIDTHS (r) OR INDIVIDUAL WIDTHS (I)? **r**

(This is identical to the option available in CPS.)

INPUT SMALLEST AND LARGEST SLOT WIDTHS (mils)? **0.0508,.3048**

The screen is cleared, and the results displayed.

PROGRAM SLOTLINE.V10 01-02-1996 22:12:33

SUBSTRATE THICKNESS = 0.160 (cm)
RELATIVE DIELECTRIC CONSTANT = 2.20
FREQUENCY = 1.575 GHz

SLOT WIDTH (cm)	Zo (ohms)	eff
0.0508	93.48	1.30
0.0762	100.44	1.28
0.1016	105.67	1.26
0.1270	109.90	1.25
0.1524	113.50	1.24
0.1778	116.69	1.22
0.2032	119.58	1.21
0.2286	122.25	1.20
0.2540	124.73	1.20
0.2794	127.07	1.19
0.3048	129.29	1.18

The user can then print out and store results if desired.

Reference [19] defines the effective length of the slot as $L_s' = L_s - W_f$. Expressions for the following parameters are also from [19]. The radiation conductance associated with the slot is given by $G_s = I/120\pi^2$ where

$$I = \int_0^{\pi} \left\{ \frac{\sin[(\pi L_s'/\lambda_0)\cos\theta]}{\cos\theta} \right\}^2 \sin^3\theta d\theta \tag{4.2}$$

which can be evaluated by standard numerical techniques. The self-inductances are found from

$$L_1 = \frac{Z_f\sqrt{\epsilon_{ref}}}{c}W_f \tag{4.3}$$

and

$$L_2 = \frac{Z_s\sqrt{\epsilon_{res}}}{c}W_s \tag{4.4}$$

where Z_f, ϵ_{ref}, and W_f are the impedance, effective dielectric constant, and width of the microstrip line; Z_s, ϵ_{res}, and W_s are the same quantities for the slotline; and c is the speed of light in vacuum (3×10^8 m/s). The mutual inductance is

$$M = \frac{\mu_0 W_s}{2\pi}\ln(\sec\,\theta_0) \tag{4.5}$$

where $\theta_0 = \tan^{-1}(L_s'/2h)$.

The open-end capacitance is found by running MICENDEF (see Appendix A) to find the end-effect extension and then using (6.49) to convert to a capacitance. PATCHCOM (see Chapter 7) will determine the radiation conductance of the open end.

Once all the circuit parameters are found, the input impedance follows from straightforward circuit analysis [19]. The total admittance of the shorted slotlines at the feed is $B_s = (-2/Z_s)[\cot(k_s L_s'/2)]$ where $k_s = 2\pi\sqrt{\epsilon_{res}}/\lambda_0$. The slot impedance, Z_s, is the parallel combination of B_s and G_s, that is, $Z_s = R_s + jX_s = 1/(G_s + jB_s)$. Finally, the input impedance is [19]

$$Z_{in} = -jZ_0\cot(k_m L_m) + \frac{M^2\omega^2}{R_s} + j[\omega(L_1 + L_2) + X_s] \tag{4.6}$$

where $k_m = 2\pi\sqrt{\epsilon_{rem}}/\lambda_0$.

To design the antenna, calculate the microstrip line width needed for the desired impedance. Estimate the slot length from the known operating frequency by assuming a free-space wavelength. Use this length to pick a slot width. Calculate the slotline parameters and then recalculate the slot length using the effective dielectric constant and accounting for the blockage of the microstrip. Evaluate C_m, G_m, B_s, G_s; then find the input impedance using (4.6). If the imaginary part is not zero at the design frequency, then change the slot length (longer if inductive, shorter if capacitive). If desired, fine tune the design by adjusting the stub length.

Some adjustment of both the slot and stub lengths will inevitably be needed to compensate for factors not included in the model. As with other designs, some experimentation is needed for the final dimensions. Either build several different combinations of slot and stub length or build one such that the lengths can be adjusted. The stub should be made longer than anticipated then trimmed back. The slot can be shortened by soldering over pieces of metal or covering with copper tape backed with conductive adhesive.

Rectangular slots (in an infinite ground plane) are electromagnetic duals of dipole antennas. They have the same radiation patterns except the *E*- and *H*-fields are reversed. A dipole radiates an *E*-field, which is parallel to the dipole axis. The slot *E*-field is perpendicular to the slot length. Diffraction effects caused by a "real world" finite-sized ground plane will cause ripples and other deviations from ideal dipole-like patterns. The slot radiates on both sides of the ground plane. To force radiation in only one direction, place the slot a quarter-wavelength above another ground plane.

Circular slots have also been constructed [21]. They are fed in the same way as rectangular slots. The radiation pattern can be varied by changing the slot radius. If the circumference is a wavelength or less, the patterns are identical to those of a wire loop antenna except for the reversal of *E*- and *H*-fields. The pattern peak is broadside to the plane of the slot. When the circumference is an integer number of wavelengths greater than one, the patterns have a null broadside to the antenna. Two pattern peaks are symmetrically located about broadside on each side of the ground plane. The angular position of the peaks varies with circumference. As the circumference increases, the peaks move away from broadside. Circular slots will radiate circular polarization if fed correctly. For a one-wavelength circumference, two feeds that are 90 degrees apart with a 90-degree phase shift between them are sufficient. For a two-wavelength antenna, four feeds are needed to suppress extraneous mode excitation. Two feeds that have a 90-degree phase shift are placed 45 degrees apart. The second pair, also with a 90-degree phase shift, are spaced 180 degrees from the first pair. The 0-degree feeds from each pair are then diametrically opposite each other as are the 90-degree feeds.

References

[1] Pozar, D. M., and B. Kaufman, "Increasing the Bandwidth of a Microstrip Antenna by Proximity Coupling," *Electronics Lett.,* April 1987, Vol. 23, No. 8, pp. 368–369.

[2] Belentepe, B., "Modeling and Design of Electromagnetically Coupled Microstrip-Patch Antennas and Arrays," *IEEE Antennas and Propagation Mag.,* Vol. 37, No. 1, Feb. 1995, pp. 31–39.

[3] Splitt, G., and M. Davidovitz, "Guidelines for Design of Electromagnetically Coupled Microstrip Patch Antennas on Two-Layer Substrates," *IEEE Trans. on Antennas and Propagation,* Vol. 38, No. 7, July 1990, pp. 1136–1140.

[4] Roy, J. S., S. K. Shaw, P. Paul, D. R. Poddar, and S. K. Chowdhury, "Some Experimental Investigations on Electromagnetically Coupled Microstrip Antennas on Two-Layer Substrates," *Microwave and Optical Tech. Lett.,* Vol. 4, No. 6, May 1991, pp. 236–238.

[5] Karmakar, N. C., and M. E. Bialkowski, "Experimental Investigations Into an Electromagnetically Coupled Microstrip Patch Antenna," *Microwave and Optical Tech. Lett.,* Vol. 5, No. 9, Aug. 1992, pp. 447–453.

[6] Pozar, D. M., "Microstrip Antenna Aperture-Coupled to a Microstripline," *Electron. Lett.,* Vol. 21, No. 2. Jan. 1985, pp. 49–50.

[7] James, J. R., and P. S. Hall, *Handbook of Microstrip Antennas,* London, UK: Peter Peregrinus Ltd., 1989, Chap. 6.

[8] Sullivan, P. L., and D. H. Schaubert, "Analysis of an Aperture Coupled Microstrip Antenna," *IEEE Trans. on Antennas and Propagation,* Vol. 34, No. 8, Aug. 1986, pp. 977–984.

[9] Ittipiboon, A., R. Oostlander, and Y. M. M. Antar, "A Modal Expansion Method of Analysis and Measurement on Aperture-Coupled Microstrip Antenna," *IEEE Trans. on Antennas and Propagation,* Vol. 39, No. 11, Nov. 1991, pp. 1567–1573.

[10] Yang, R., Y. T. Lo, M. I. Aksun, and S. L. Chuang, "Simple and Efficient Analysis for a Slot-Coupled Patch Antenna with a Microstrip Line Feed," *Microwave and Optical Tech. Lett.,* Vol. 4, No. 9, Aug. 1991, pp. 335–341.

[11] Wen, C. P., "Coplanar Waveguide: A Surface Strip Transmission Line Suitable for Non-Reciprocal Gyromagnetic Device Application," *IEEE Trans. on Microwave Theory and Techniques,* Vol. 17, No. 12, Dec. 1969, pp. 1087–1090.

[12] Gupta, K. C., R. Garg, and I. J. Bahl, *Microstrip Lines and Slotlines,* Norwood, MA: Artech House, Inc., 1979, Chap. 7.

[13] Chang, K., ed., *Handbook of Microwave and Optical Components,* Vol. 1, New York, NY: John Wiley & Sons, 1989, pp. 34–36.

[14] Menzel, W., and W. Grabherr, "A Microstrip Patch Antenna With Coplanar Feed Line," *IEEE Microwave and Guided Wave Lett.,* Vol. 1, No. 11, Nov., 1991, pp. 340–342.

[15] Smith, R. L., and J. T. Williams, "Coplanar Waveguide Feed for Microstrip Patch Antennas," *Electron. Lett.,* Vol. 28, No. 25, Dec. 1992, pp. 2272–2274.

[16] Edward, B., and D. Reese, "A Broadband Printed Dipole with Integrated Balun," *Microwave J.,* Vol. 30, No. 5, May 1987, pp. 339–344.

[17] Balanis, C. A., *Antenna Theory Analysis and Design,* New York, NY: Harper & Row, 1982, Chap. 7.

[18] Levine, E., S. Shtrikman, and D. Treves, "Double-Sided Printed Arrays with Large Bandwidth," *IEE Proc. Pt. H (Microwaves, Optics and Acoustics),* Vol. 135, No. 1, Feb. 1988, pp. 54–59.

[19] Akhavan, H. G., and D. Mirshekar-Syahkal, "Approximate Model for Microstrip Fed Slot Antennas," *Electron. Lett.,* Vol. 30, No. 23, Nov. 1994, pp. 1902–1903.

[20] Janaswamy, R., and D. H. Schaubert, "Characteristic Impedance of a Wide Slotline on Low-Permittivity Substrates," *IEEE Trans. on Microwave Theory and Techniques,* Vol. 34, No. 8, Aug. 1986, pp. 900–902.

[21] Bahl, I. J., and P. Bhartia, *Microstrip Antennas,* Norwood, MA: Artech House, Inc., 1980, pp. 238–242.

Chapter 5

Circularly Polarized Element Design

The microstrip radiators covered in Chapter 3 are all linearly polarized. In many instances a circularly polarized antenna is desirable. For example, it may be difficult to know beforehand the required orientation of the antenna when linear polarization is used. Satellite-to-mobile ground-based or airborne links are a good example of this problem. In other situations, a circularly polarized antenna may make a system more user-friendly by avoiding the need to line-up the antenna with the signal polarization. Right- and left-hand circular polarizations are orthogonal to each other. This can be employed to double the channel capacity on a link by having one signal use right-hand and the other left circular polarization. The GPS antenna that has been used in the design examples in actuality needs to be a circularly polarized antenna. This not only eliminates the need to orient the antenna but also maximizes the received signal and can reduce the effects of multipath. Two techniques for designing circularly polarized microstrip antennas are covered in this chapter. One uses a square patch designed by the methods of Chapter 3. The circular polarization is obtained by using two feeds that have a specified phase and amplitude relationship. The second modifies the shape of a square patch to produce an antenna that radiates circular polarization with only one feed.

5.1 CPPATCH

This chapter describes the design of circularly polarized patch antennas. All of the programs mentioned in the chapter but one have already been introduced in other chapters. The only new program is CPPATCH.

CPPATCH: Designs a single point-fed circularly polarized patch. Does either a truncated or nearly square patch. Combines the curve-fit formulas of PATCHD with a perturbation analysis of a modified cavity to analyze the antenna. Computes resonant length and impedance as a function of frequency. Includes effect of feed probe.

5.2 SINGLE-FEED CIRCULARLY POLARIZED ELEMENT

A circularly polarized wave has two orthogonal field components that are in phase quadrature. The dual-fed patch, to be discussed in Section 5.4, uses two feeds to excite two spatially orthogonal, resonant modes, which produces a circularly polarized wave. An external feed network ensures that the modes have equal amplitudes and are in phase quadrature. Microstrip antennas can be designed to radiate circular polarization with no external network and only one feed. It is possible to excite the two modes with one feed by introducing a small perturbation in the patch shape [1]. Many types of perturbations are used. Two of the most common are shown in Figure 5.1. In Figure 5.1(a), two diagonally opposite corners are truncated much like the chamfer on a microstrip bend. The patch is square except for the corners and is fed at the center of one of its sides. The antenna is fed along a center line as with a linearly polarized patch. If fed from the side indicated on the figure, the polarization will be right-hand circular. When fed from an adjacent side, it will be left-hand circular. The antenna of Figure 5.1(b) has one side slightly longer than the other forming a nearly square patch. These antennas are fed either in the corner or along the corresponding diagonal. As shown, it radiates right-hand circular polarization. Left-hand circular occurs when the feed is placed on the other diagonal.

The feed excites fields under the patch just as with a linearly polarized antenna. The signal injected by the feed tends to propagate in one direction, guided by the transmission line formed by the patch. In very approximate terms, the perturbation segment "scatters" the feed fields into a mode that is spatially orthogonal. Since this scattered mode "sees" a different patch geometry, its resonant frequency is shifted slightly. To achieve circular polarization, these two modes must be made equal in amplitude and must differ in phase by 90 degrees. The equal amplitude is obtained by proper positioning of the feed. This is why the feed is placed along the patch center line for the truncated

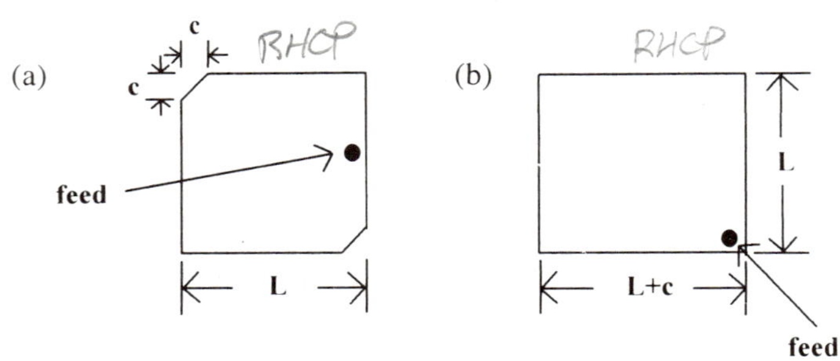

Figure 5.1 (a) Truncated and (b) nearly square single-feed circularly polarized patches.

patch and along a diagonal for the nearly square patch. The creation of a 90-degree phase shift is due to two factors. One is again the feed position, but this time it involves the orientation of the feed with respect to the perturbation. Shifting from one side to the adjacent side changes the sense of the polarization for the truncated patch. The other factor is the size of the perturbation.

Reference [1] analyzes the antennas of Figure 5.1 using a variational technique similar to that used for covered microstrip in Section 3.8. Assume a square patch lying in the (x-y)-plane with its length parallel to the y-axis. When a square patch is fed with two feeds, one along its width and one along its length, modes are excited that are resonant along the x- and y-directions. For the purposes of this analysis, consider only the dominant mode in each direction. From the cavity model, the vertical electric field under the patch for the mode resonant along the patch length is

$$E_z^l \propto E_0 \cos(k_m y) \tag{5.1}$$

while that for the orthogonal mode is

$$E_z^w \propto E_0 \cos(k_n x) \tag{5.2}$$

For the singly-fed patch, only one mode, say (5.1), is excited by the feed. The perturbation scatters some of this into a mode similar to (5.2). In actuality, the perturbation also modifies the original mode, creating two new modes. In the case of the truncated patch, the new modes are not resonant along the x- and y-axes. Expressions like (5.1) and (5.2) still hold, however.

The new modes have new functional dependencies and values for k_n and k_m. Since the perturbation is small, the changes in these quantities are expected to be small as well. Under these conditions, the new modes can be expressed as a linear combination of the old modes, (5.1) and (5.2), with unknown weightings. By applying variational theory, an integral expression can be derived for the new k values. From the new k's, the resonant frequencies of the modes are found. A normalization procedure is applied to the linear combination of the old modes to determine the weighting coefficients. From this analysis it is possible to derive the conditions for circular polarization and determine an equivalent circuit for the patch.

The results of the analysis determine the conditions for circular polarization and identify the required locations for the feed as discussed earlier. They also result in an expression for the size of the perturbation, which turns out to be related to the antenna Q [1]. For the truncated patch

$$\frac{\Delta s}{S} = \frac{1}{2Q} \tag{5.3}$$

and for the nearly square patch

$$\frac{\Delta s}{S} = \frac{1}{Q} \tag{5.4}$$

where Δs is the area of the perturbation segment, S is the area of the patch, and Q is the patch quality factor. If the starting square patch length is denoted by L, then $S = L^2$ and

$$c = \sqrt{\Delta s} \tag{5.5}$$

for the truncated patch, while with the nearly square patch

$$c = \frac{\Delta s}{L} \tag{5.6}$$

c is either the length of the side of the cut-out triangle or the length extension of one side of the patch.

Let the resonant frequencies for the new modes be denoted by f_a and f_b. These are functions of the perturbation segment area. They are first for the truncated and then the nearly square antennas, respectively,

$$f_a = f_0\left[1 - \frac{2\ \Delta s}{S}\right] \tag{5.7}$$

$$f_b = f_0 \tag{5.8}$$

$$f_a = f_0\left[1 - \frac{\Delta s}{S}\right] \tag{5.9}$$

$$f_b = f_0 \tag{5.10}$$

where f_0 is the resonant frequency of the unperturbed antenna. The conditions for circular polarization occur at the arithmetic mean of the resonant frequencies. Note that one of the new frequencies is identical to the square patch frequency. To design for a given frequency, the starting square patch should be slightly smaller than for a corresponding linearly polarized one.

At the design frequency, the two resonant modes are dominant. Near the resonant frequencies, each mode can be modeled by a parallel RLC circuit. The resistance represents the losses associated with the mode, which are primarily due to radiation. The inductance and capacitance model the energy storage. The modes are in series, producing the equivalent circuit given in Figure 5.2. Each mode has a transformer at its input, representing the amount of excitation of that mode. When the circular polarization conditions are

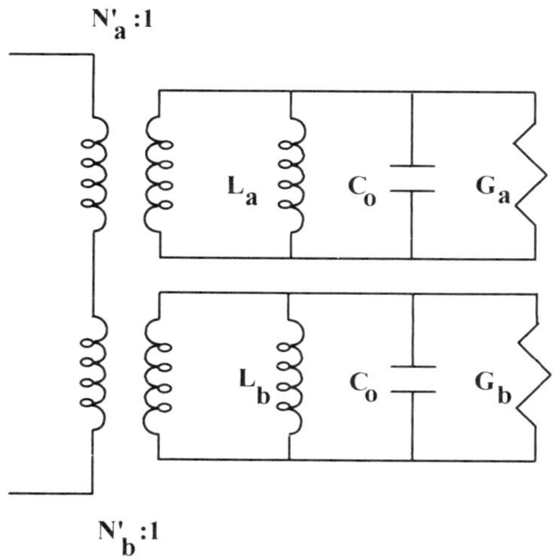

Figure 5.2 Equivalent circuit for a single-fed circular polarized patch.

satisfied, the turns ratio for each are equal (for example, $N'_a = N'_b$). The capacitance is given by (3.12). The conductances are found using (3.11). For all practical purposes, the mode Q's are the same; therefore, the conductances are identical. The inductance is found by setting the reactance to zero at the mode resonant frequency, $L_i = 1/(\omega_i^2 C_0)$ where $i = a, b$.

The turns ratios, which are equal to each other at resonance, are a function of feed position and frequency. Let the center of the unperturbed patch be the origin of an x-y coordinate system as in Figure 5.3. For the truncated patch, the feed lies along the x- or y-axes. The feed lies along the diagonals for the nearly square patch. The expressions for the truncated patch transformers are [1]

$$N'_a = (\sqrt{S}/L)[\sin(kx) - \sin(ky)] \tag{5.11}$$

and

$$N'_b = (\sqrt{S}/L)[\sin(kx) + \sin(ky)] \tag{5.12}$$

The corresponding expressions for the nearly square patch are [1]

$$N'_a = \sqrt{2}\,\sin(kx) \tag{5.13}$$

and

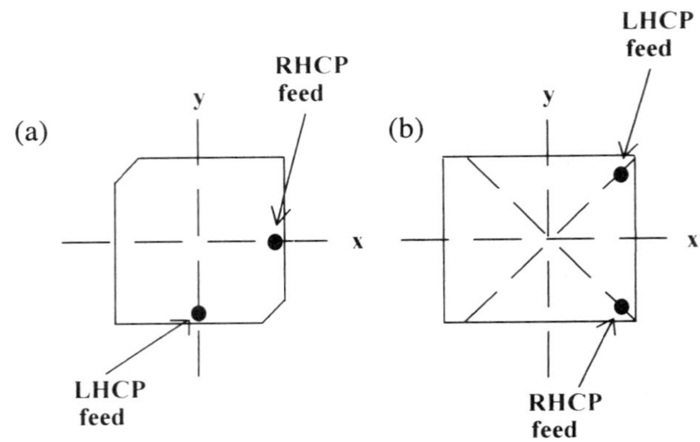

Figure 5.3 Single-feed patches with coordinates for feed locations: (a) truncated patch and (b) nearly square patch.

$$N_b' = \sqrt{2}\,\sin(ky) \tag{5.14}$$

where $k = \pi/L$.

5.3 SINGLE-FEED PATCH DESIGN

Program CPPATCH implements the above theory for the design of single-feed circularly polarized patches. CPPATCH first finds the resonant length, Q and C_0 of the corresponding square patch using the curve-fit formulas of PATCHD. From the patch Q and length, the truncation dimensions, mode resonant frequencies, and mode circuit elements are determined using (5.3) to (5.14). CPPATCH then calculates the impedance versus frequency of the circuit in Figure 5.2. The feed can be inset, and probe circuit elements can be included. The probe model is the same as in Section 3.3 (see Figure 3.3). Results are displayed on the screen both in tabular form and on a Smith chart. They can also printed out or stored in a file.

A design example with a 1.575-GHz patch will serve to show how the program is used. Executing CPPATCH, the screen looks as follows.

CPPATCH.V20 8/17/95 13:05:34
[Current Program Date - 17/8/95]

Circularly Polarized Patch Design

DO NEARLY SQUARE (n) OR TRUNCATED (t) PATCH DESIGN? **n**

RIGHT (r) OR LEFT (l) CIRCULAR POLARIZATION? **r**

INPUT ANTENNA FREQUENCY (GHz)? **1.575**

INPUT SUBSTRATE RELATIVE DIELECTRIC CONSTANT AND LOSS TANGENT? **2.2,.001**

INPUT SUBSTRATE HEIGHT (cm)? **.16**

INPUT CONDUCTOR CONDUCTIVITY RELATIVE TO COPPER
(default value is 1)
?<rtn>.

CPPATCH then calculates the patch dimension and Q. For the nearly square patch, the next input is the feed position

INPUT FEED INSET (cm) ALONG DIAGONAL
[0 <= yf < 4.653]
[patch center   patch edge]
?1.32

The *4.653* dimension is the calculated patch length multiplied by $\sqrt{2}$, which gives the length along the diagonal. If a truncated patch is being designed, the program would ask for the feed position along the patch center line. The next input is

INCLUDE FEED PROBE? **n**

If answered yes (**y**), a request is made for the feed probe and relief hole diameters. As with other programs, some inputs are tested against model limits, and if the limits are exceeded, the user is given the opportunity to change the inputs. CPPATCH then asks for frequency data

INPUT START AND STOP FREQUENCIES (GHz)? **1.525,1.625**

Like most of the models in this book, CPPATCH is only valid near the patch resonance, so the frequency range should be restricted to be within a few percent.

INPUT NUMBER OF FREQUENCIES TO CALCULATE IMPEDANCE? **20**

The final calculation is the evaluation of the circuit model of Figure 5.2 to obtain the impedance as a function of frequency.

When that calculation is completed, the screen is cleared and the following is displayed.

CPPATCH.V20   11-05-1995   11:27:42

RHCP NEARLY SQUARE PATCH

SUBSTRATE HEIGHT = 0.160 cm
SUBSTRATE RELATIVE DIELECTRIC CONSTANT = 2.20
SUBSTRATE LOSS TANGENT = 0.0010
CONDUCTOR RELATIVE CONDUCTIVITY = 1.000
NOMINAL PATCH SIZE = 6.261 cm
FEED LOCATION = 4.110,–4.064
WIDTH EXTENSION = 0.094 (cm)
FREQUENCY = 1.5736 GHz

FREQ (GHz)	*PATCH IMPEDANCE (ohms)*
1.525	*26.62 j 70.58*
1.530	*36.73 j 79.43*
1.535	*52.99 j 88.60*
1.540	*79.12 j 93.94*
1.545	*114.98 j 83.60*
1.550	*140.16 j 46.71*
1.555	*132.27 j 9.89*
1.560	*117.17 j −0.79*
1.565	*119.64 j −1.78*
1.570	*135.67 j −18.73*
1.575	*135.24 j −58.58*
1.580	*105.00 j −88.27*
1.585	*71.83 j −93.22*
1.590	*48.97 j −86.67*
1.595	*34.69 j −77.79*
1.600	*25.65 j −69.51*
1.605	*19.68 j −62.41*
1.610	*15.57 j −56.46*
1.615	*12.64 j −51.48*
1.620	*10.46 j −47.27*
1.625	*8.82 j −43.68*

The frequency where all the conditions are met for circular polarization is printed just above the impedance data. It is slightly below the input design frequency. The (x,y) coordinates of the feed (see Figure 5.3 for the location of the origin) are printed out just below the nominal patch size. The width extension, which is the perturbation, is printed next.

PLOT OUT RESULTS ON A SMITH CHART? **y**

Figure 5.4 shows the Smith chart plot for this case. Note that the impedance curve has a kink in it. The tip of the kink is in the region of best polarization performance. The shape of the curve is a result of having two modes with slightly different resonant frequencies. Next the user is given the option to print out or store data.

PRINT OUT RESULTS (y or n)? **n**

STORE RESULTS IN A FILE (y or n)? **n**

The output can be stored in a file that essentially duplicates the screen output or in a TOUCHSTONE compatible file format. The final input is

DO ANOTHER CASE (y or n)? **n**

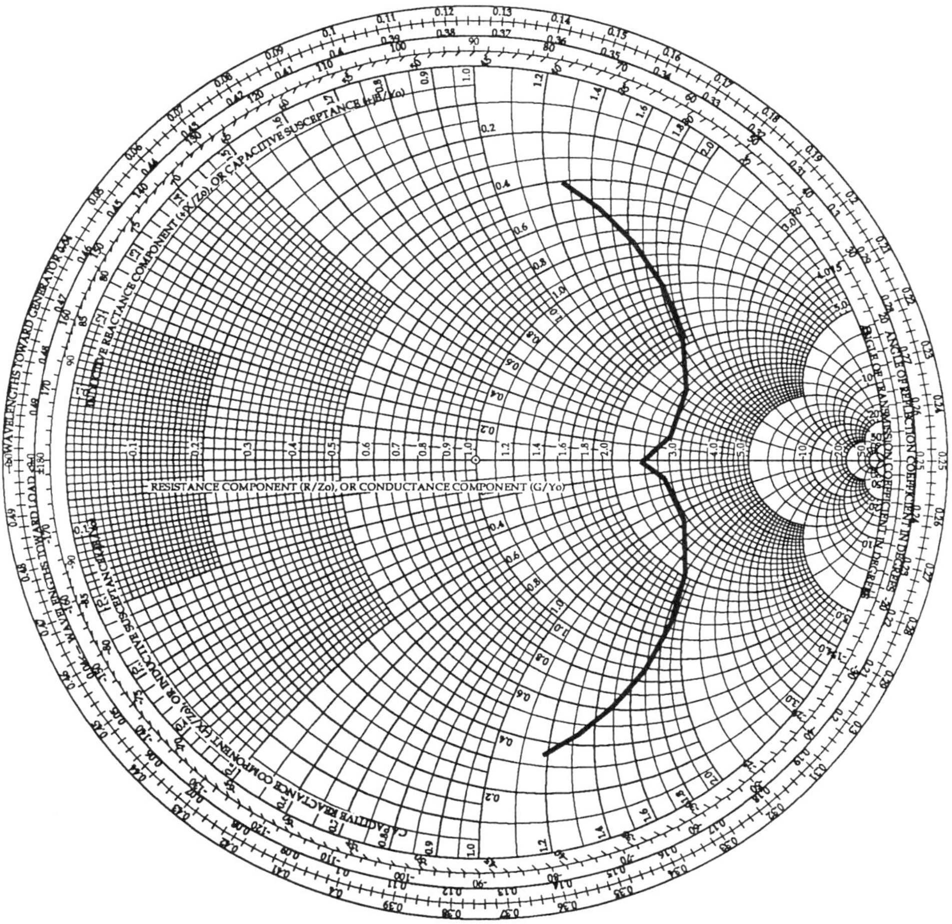

Figure 5.4 Impedance versus frequency for the nearly square patch design example.

which completes the program run.

The model incorporated in CPPATCH is based upon an approximate analysis. The results are probably accurate to within a few percent. While this seems a reasonable accuracy, the bandwidth for good circular polarization behavior is very small, usually only about 1%. Thus the model predictions will be useful for establishing starting dimensions, but additional experimental work may be needed to obtain the final design.

Some results for several single point-feed antennas are presented in [2]. Both theoretical and experimental data are given. The theoretical approach employed is different from that described here and is probably somewhat more accurate. The best axial ratio occurs when the feed is at the patch edge. Unfortunately the impedance is highest at this point, producing a high SWR with respect to a 50-Ω feed. The SWR (and impedance) is lowest at a point about 0.3 times the distance to the edge as measured from the patch center. The minimum SWR is determined by the substrate thickness. Using thinner substrates decreases the impedance, so it is possible to find a thickness where the minimum impedance is close to 50Ω. The antenna bandwidth is determined solely by the frequency range over which reasonable circular polarization is maintained. It can be shown [3] that the bandwidth over which the axial ratio at the pattern peak is 3 dB or less is about $35/Q$. Unfortunately the polarization bandwidth increases with the substrate thickness, which is contrary to what is required for a good impedance match. The performance of the antenna is relatively independent of the type of perturbation.

5.4 DUAL-FEED CIRCULARLY POLARIZED ELEMENT

As discussed in Chapter 1, a circularly polarized antenna radiates a far field that contains two electric field components. They are spatially orthogonal to each other and in phase quadrature (for example, one component is delayed, or advanced, by 90 degrees). In the following only rectangular patches will be considered, but most of what is discussed is equally applicable to circular patches. The rectangular patch is linearly polarized. According to the transmission line model, the feed injects a signal into the patch. The patch acts as a transmission line, and the signal travels along the length of the patch. When it reaches the edges, the discontinuity caused by the truncation of the metalization results in radiation. The patch length is approximately a half-wavelength long. This not only makes the input impedance purely real but also causes the radiation from the ends to be in phase broadside to the patch. The far-field pattern thus has a maximum perpendicular to the patch surface.

The patch width can be chosen to be less than, equal to, or greater than the length. The width controls the input resistance, and, to a lesser degree, the width effects the *H*-plane pattern. Typical widths are from 0.5 to 2.0 times the length. One reason why the resonant frequency is set by the patch length is because of the feed. The orientation of the feed and the way it excites fields underneath the patch favors the propagation of the signal along the patch length. If the feed were to be placed in the center of an edge along the patch length, the patch would be resonant at a frequency determined by the width. The patch thus has two potential resonant frequencies—one determined by the length and the other by the width.

If the patch is made square, it will have two resonant frequencies that are identical. The far fields from each will be spatially orthogonal to each other. If they can be set-up to have equal amplitudes and a 90-degree phase difference, then the patch will radiate a circularly polarized signal. The easiest way to do this is to have two feeds—one along the width and one along the length. The amplitude and phase relationships are then established by a network that is external to the antenna. A square patch with two correctly excited, orthogonal feeds is a popular technique for generating circular polarization.

Note that essentially the same argument applies to a circular patch. Since there is no "length," a resonance is established for a feed along any radial line contained in the patch. Again two spatially orthogonal feeds will produce orthogonal far-field components. These will generate circular polarization if phased correctly.

There are two common realizations for the network that establishes the correct amplitude and phase relationships for the feeds. These are shown in Figure 5.5. The patch in Figure 5.5(a) is fed with a circuit known as a quadrature hybrid, which is a type of directional coupler. The other feed type is simply a microstrip T-junction where one of the arms that splits off to the antenna is made 90 degrees longer than the other as in Figure 5.5(b). This circuit is sometimes called a reactive splitter because it is not an inherently impedance-matched device. Depending on the actual implementation, quarterwave impedance transformers (not shown in the figure) may be used. These would be placed at each antenna feed and connect the antenna to the hybrid or splitter network. The impedance of a square patch is quite high, so the transformer may be needed to bring the impedance level down to a value more suitable for the networks.

The choice of which feed to use depends on several factors, including electrical performance, size constraints, and cost. The hybrid feed offers better electrical performance. It will provide the correct excitations over a broader bandwidth. The outputs are isolated from each other, so variations in antenna impedance or asymmetries in the circuit do not affect the excitations. This is not the case for the splitter, which is not isolated. The hybrid is matched at all ports. The T-junction splitter, on the other hand, takes up considerably less space. It is also lower in cost because the unused port of the hybrid must be terminated in a matched load, typically a 50-Ω resistor. For many applications, the splitter's poorer performance can be tolerated for the savings in space and cost.

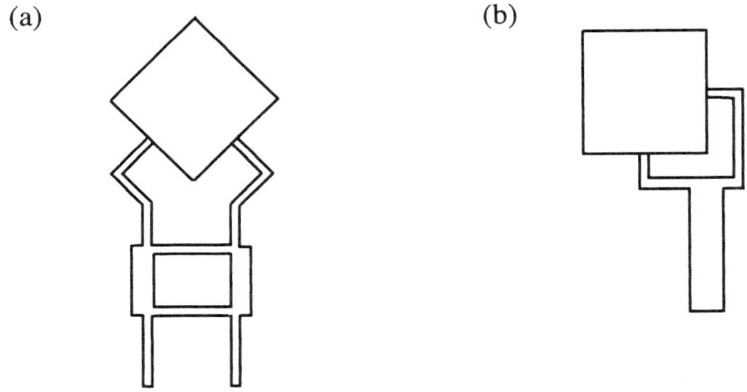

Figure 5.5 Two-feed networks for circularly polarized square microstrip patches: (a) hybrid-fed patch and (b) reactive splitter-fed patch.

5.5 QUADRATURE HYBRID FEED

The quadrature hybrid or branch line coupler is a widely used microstrip circuit. Most commercially available hybrids are realized in microstrip or stripline. The configuration of a microstrip hybrid is shown in Figure 5.6. A quadrature hybrid is a four-port device; that is, it has four terminals. To form the hybrid, four microstrip lines are arranged in a square. All lines are a quarter-wavelength long. The series arms have a characteristic impedance of Z_f, while the shunt arm impedance is Z_p. In Figure 5.6, all four terminals are connected to transmission lines of impedance Z_0, which is the reference impedance of the system, usually 50Ω.

When the hybrid is connected to matched loads at all the ports, a signal incident on port 1 will be coupled to ports 2 and 3. No signal will appear at port 4. The signal at port 3 lags that of port 2 by a phase difference of 90 degrees. The magnitudes of the signals at 2 and 3 depend upon the ratio of the shunt and series arm impedances. Usually they are chosen so that the signals are equal. Note that the hybrid is completely symmetrical, and any port can be used as an input. If a signal is input to port 4; it will split between 2 and 3 with port 1 being isolated. If port 3 lags port 2 by 90 degrees when 1 is the input, it will lead by 90 degrees for an input at 4.

The design equations for the hybrid are [4]

$$\frac{P_2}{P_3} = \left(\frac{Z_0}{Z_p}\right)^2 \tag{5.15}$$

and

$$\frac{Z_0}{Z_f} = \left[\left(\frac{Z_0}{Z_p}\right)^2 + 1\right]^{1/2} \tag{5.16}$$

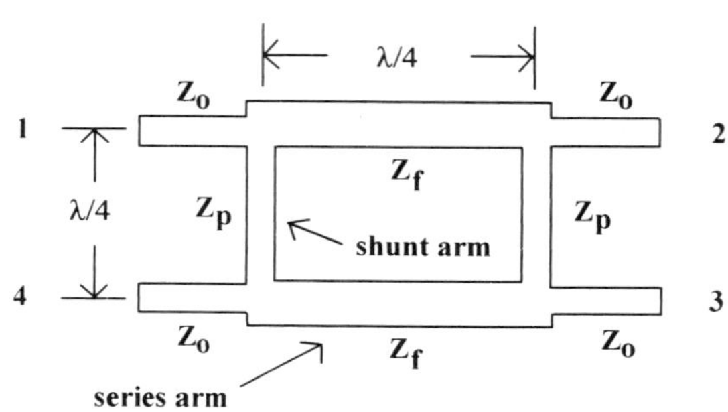

Figure 5.6 Microstrip quadrature hybrid configuration.

where P_2 and P_3 are the powers delivered to ports 2 and 3. For an equal power split, which is the usual case, $Z_p = Z_0$ and $Z_f = Z_0/\sqrt{2}$. The line lengths are found by first determining the effective dielectric constant of the arms and then calculating the wavelength at the operating frequency. Because of the parasitics associated with the T-junctions, the arms are made slightly longer than a quarter-wavelength. The series arm is longer by about twice the reference plane extension, d_1; while the shunt arm is increased by about twice the extension, d_2 (see Figure 2.3 for definitions of the extensions). Hybrids have a bandwidth of about 10%.

It is not necessary for all four arms to operate into the same reference impedances. In fact, hybrids can perform an impedance-matching function as well [5]. Assume the impedance level at ports 1 and 4 is still Z_0, but the impedance of the loads at 2 and 3 is Z_d. The design equations given in [5] are expressed in terms of admittances. Let Y_d be the admittance of the output port load normalized to the input port admittance, Y_0. The characteristic admittance of the series arms is

$$Y_f = \sqrt{\frac{(K+1)Y_d}{K}}Y_0 \tag{5.17}$$

where $K = P_2/P_3$ is the power split between outputs, $Y_d = Z_0/Z_d$ and $Y_0 = 1/Z_0$. The shunt arm impedances are no longer identical. The shunt arm adjacent to the input side is given by

$$Y_p = \frac{Y_0}{\sqrt{K}} \tag{5.18}$$

while the shunt arm at the output side is

$$Y_p = \sqrt{\frac{Y_d^2}{K}}Y_0 \tag{5.19}$$

This impedance matching property can sometimes be useful for reducing circuit area by eliminating the normally used matching circuit.

Hybrids are isolated devices; that is, there is no interaction between output ports. This is a desirable property for antenna feeds because when the outputs are not isolated, they can interact with each other and disrupt the excitation. Examples of this occur when the patches are not matched or the impedance is different from patch to patch. These are very common occurrences. When properly terminated in matched loads, the hybrid is matched at all ports. Another interesting property of hybrids can be seen when the output loads are not matched. If both loads are identically mismatched, their reflected signals appear at the isolated port. The input port is still matched. This means that the isolated port should be terminated to absorb the reflected power. Although the input is matched, power is lost into the isolated port termination. A reduction in antenna gain occurs and the efficiency is also reduced.

The area occupied by a hybrid is just over a quarter-wavelength square. At low frequencies, especially below 1 GHz, this becomes quite large. Hybrids can be constructed in lumped element form [6,7] which results in a considerable space saving. The lumped elements are, of course, quite readily realizable at low frequencies. Even at higher frequencies it may be desirable to use a lumped-element hybrid if it is to be placed on a chip.

To realize the hybrid, the various transmission line sections are transformed into high-pass equivalent π-networks [6]. The resulting equivalent circuit can be seen in Figure 5.7. The capacitances and inductances are obtained from the image parameters of the π-networks. Let the design frequency be f_0. The capacitances are found from

$$C_{ij} = \frac{1}{2\pi f_0 Z_{0i}} \tag{5.20}$$

where

$$Z_{0i} = \begin{cases} Z_f & \text{for } C_{12} \\ Z_p & \text{for } C_{14} \end{cases}$$

while the inductance is

$$L_{t1} = \frac{Z_f Z_p}{2\ \pi f_0 (Z_f + Z_p)} \tag{5.21}$$

These equations are for equal power split hybrids. See [6] for hybrids with an unequal power split between outputs.

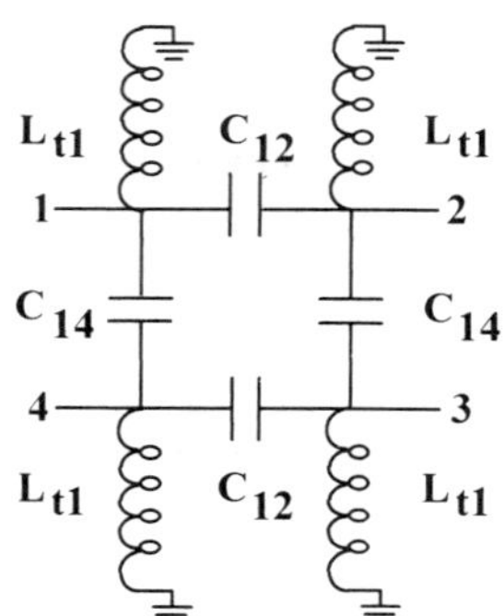

Figure 5.7 High-pass lumped-element equivalent circuit for a quadrature hybrid.

A low-pass version of the lumped hybrid can also be designed [7]. In the equivalent circuit of Figure 5.7, L_{t1} is replaced by a capacitor C_0, C_{12} is replaced by an inductor L, and C_{14} is replaced by another capacitor C_1. The design equations given in [7] are

$$C_1 = \frac{1}{2\ \pi f_0 Z_0 \sqrt{K}} \tag{5.22}$$

$$(2\ \pi f_0)^2 L(C_0 - C_1) = 1 \tag{5.23}$$

and

$$L = \frac{2 C_0 Z_0^2}{1 + (2\ \pi f_0 C_0 Z_0)^2} \tag{5.24}$$

where Z_0 and K are as previously defined. Equations (5.22) to (5.24) are solved by a bisectional method. Due to the nature of these equations, a solution for L and C_0 is guaranteed. The solution starts with three initial guesses for C_0 (for example, 0 pF, 500 pF, and 1000 pF). For each guessed value of C_0, a value of L is found from both (5.23) and (5.24). The difference in the two L values is a monotonically decreasing function of C_0. The solution lies between the values of C_0, where the difference in L changes sign. A computer program that solves the equations can be found in [7].

5.6 CIRCULARLY POLARIZED PATCH DESIGN USING A HYBRID

The design begins by determining the patch size and impedance. Initially assume that the antenna will be fed with microstrip lines on the same surface as the patch (Figure 5.5(a)). PATCH9 is executed to design the patch. The patch length is not known beforehand, so an iterative process is used to find the width. One of the inputs is the patch length. PATCH9 provides an estimated length based upon $\lambda_0/(2\sqrt{\epsilon_r})$. In lieu of a better value, this is usually a good starting point for the calculations. After entering the PATCH9 estimated length, enter the same value for the next input, which is the width. PATCH9 will use the model to determine a more accurate length. Once the calculations are complete, rerun PATCH9 using the updated length as the patch length and width. Usually the length does not change significantly during this second calculation, so the output will be the results for the required square patch.

Let the design frequency be 1.575 GHz. The substrate is chosen to be 0.16-cm thick with a dielectric constant of 2.2 and loss tangent of 0.001. The metalization is copper with a thickness of 0.00356 cm (1-oz. copper). After running PATCH9 twice, the patch dimensions are found to be 6.302 cm by 6.302 cm, and the impedance at resonance is 281Ω with no feed inset. With microstrip feeds, it is advisable to use either no inset or a very slight one. The cut-out portion of the patch for the inset of one feed may effect the fields of the orthogonal feed and cause a degradation in the circular polarization. Continuing with the design, the interface between the patch and the hybrid needs to be considered because of the high patch impedance. Two options are available: One is to quarterwave transform the patch impedance to a more suitable value such as 50Ω, while the other is to design the impedance matching into the hybrid.

For the quarterwave transformer approach, its characteristic impedance is

$$Z_t = \sqrt{Z_0 Z_a} = \sqrt{50(281)} = 118.5\Omega \tag{5.25}$$

with Z_a being the patch impedance. The main concern for the transformer is the realizability of the impedance. High impedances require narrow line widths, and the impedance becomes a sensitive function of the width for narrow lines. Not only is the accuracy of the line width important, but the width has to be kept very constant along the line length. Using MICRO, a 118.5-Ω line requires a 0.0899-cm-wide line, which is readily achievable. The effective dielectric constant is $\epsilon_{re}(f) = 1.726$, giving a transformer length of 3.624 cm.

The next step is the hybrid design. When using transformers, the impedance at all ports is 50Ω. For an equal power split between outputs, the series arm impedance is $Z_0/\sqrt{2} = 35.4\Omega$, while the shunt arms are equal to the port impedance of 50Ω. The series arm width and effective dielectric constant (from MICRO) are 0.806 cm and 1.933 cm, respectively. The corresponding values for the shunt arm are 0.495 cm and 1.875 cm. The nominal lengths are 3.424 cm and 3.477 cm. To compensate for the junction effects, MICTEE is run for a junction with a 50-Ω main line entering from the left, a 35-Ω main line entering from the right, and a 50-Ω line branch line. The reference plane extension for the 35-Ω side is about 0.0328 cm. This is added to each side of the series arm, making its length 3.490 cm. On the branch line, the extension is 0.287 cm, resulting in a shunt arm length of 4.051 cm. The hybrid design is shown in Figure 5.8. The line lengths are measured from the center of one junction to the center of the other junction.

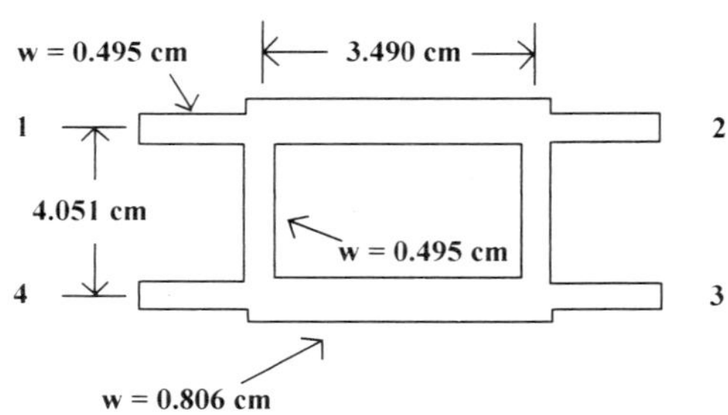

Figure 5.8 Hybrid design for 1.575-GHz circularly polarized patch.

The line lengths are "best guesses." It is probably good practice to experimentally check the design and perform any needed corrections to the lengths.

The final design is shown in Figure 5.9. The transformer lines may have to be bent as in the figure. If the angle of the bend is small (for example, less than about 30 degrees), there is no need to use a chamfer at the corner. When the antenna is fed from port 1, the radiated fields will be right-hand circularly polarized. Conversely, left-hand polarization occurs when port 4 is used. Whichever port is not fed should be terminated in a matched (50-Ω) load. See Figure 7.16 and accompanying discussion for techniques to realize a 50-Ω load. To maintain the phase difference, it is important to keep the two line lengths from the antenna to the hybrid equal. The bandwidth of this antenna is about 4% to 6%.

The other design option when using a hybrid network is to perform the impedance matching within the hybrid. A hybrid can be designed to have output port impedances of 279Ω using (5.17) to (5.19). To realize the hybrid, the shunt line on the output side has to have a characteristic impedance of 281Ω (see (5.19)). The required line width is 0.000864 cm. The line is too narrow to fabricate without special processing procedures, which increases cost considerably. The hybrid could be designed for another output impedance level, but this would necessitate quarterwave transformers at the antenna. This offers no advantage to the previous design. Using an impedance matching hybrid does not appear to be a viable approach for this application.

5.7 T-JUNCTION FEED

Circularly polarized patches can also be fed with a T-junction splitter as in Figure 5.5(b). Figure 5.10 presents a general layout for a microstrip splitter. In general, the splitter will have three lines with differing characteristic impedances. A simplified circuit model for the splitter, commonly used for design purposes, is shown in Figure 5.11 [8]. There are three transmission lines connected to a common point. The susceptance, B, represents

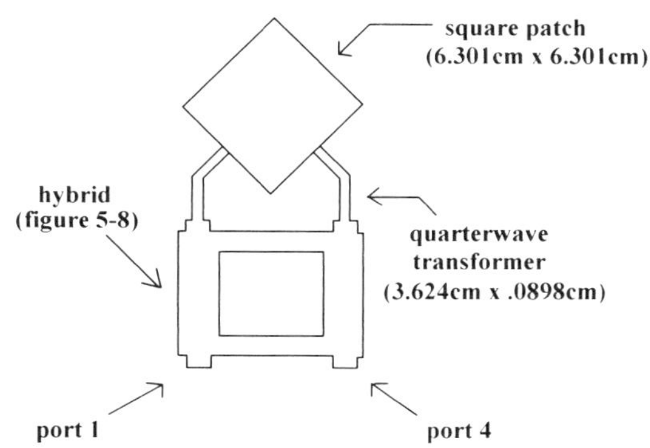

Figure 5.9 Design of a hybrid-fed circularly polarized patch.

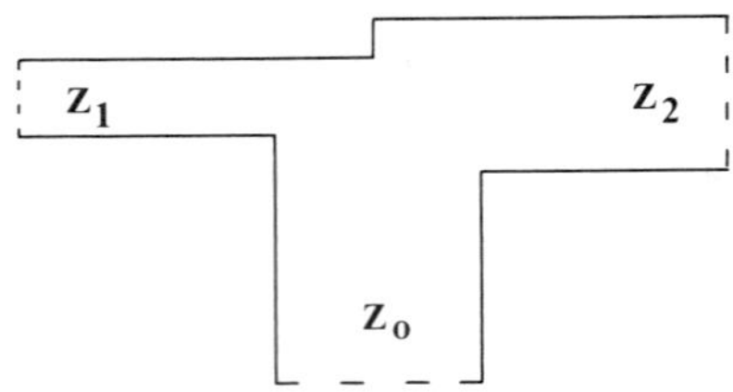

Figure 5.10 Microstrip T-junction splitter.

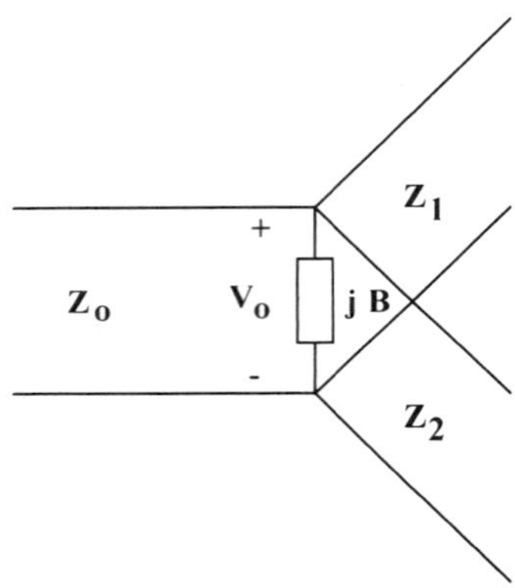

Figure 5.11 T-junction splitter simplified equivalent circuit.

energy stored in the junction due to parasitics. Other than B, the splitter model does not contain the other parts of the microstrip T-junction model. The power split between outputs is determined by the ratio of the impedances, Z_1 and Z_2.

As an example, assume that a 3:1 split is required. The power input to the splitter is

$$P_{\text{in}} = \frac{1}{2}\frac{V_0^2}{Z_0} \tag{5.26}$$

For the desired output power split,

$$P_1 = \frac{1}{2}\frac{V_0^2}{Z_1} = \frac{1}{4}P_{\text{in}} \tag{5.27}$$

and

$$P_2 = \frac{1}{2}\frac{V_0^2}{Z_2} = \frac{3}{4}P_{in} \tag{5.28}$$

Substituting (5.26) into (5.27) and (5.28), the impedances of the output lines are found to be

$$Z_1 = 4Z_0 \tag{5.29}$$

and

$$Z_2 = \frac{4}{3}Z_0 \tag{5.30}$$

If the junction susceptance, B, is negligible, the input impedance of the splitter is the parallel combination of the output lines, which is Z_0.

Within the validity of the model and assuming that $B = 0$, the splitter has a matched input; that is, it is equal to the characteristic impedance of the input line, Z_0. The outputs are not matched, however. For the Z_1 arm, the output impedance is $4Z_0/5$. The other arm impedance is $4Z_0/7$. If the loads on the outputs are not these values, the match can be improved by using quarterwave transformers at the outputs. A three-terminal device like the T-junction splitter can never be perfectly matched at all ports. For most applications, no attempt is made to improve the match of the outputs. Since the outputs are not isolated, the loads need to be the correct impedances in order for the splitter to work properly. This is an important point with unmatched, nonisolated circuits. It is also good practice to exploit symmetry where possible.

In some cases the impedances determined from (5.29) and (5.30) may not be suitable for implementation in microstrip. The impedances can be either too high or too low. One solution is to calculate Z_1 and Z_2 using an intermediary value of impedance instead of Z_0. Let this impedance be Z_e. The splitter input impedance is then Z_e, which can be matched (to Z_0) with another transformer of impedance $\sqrt{Z_0 Z_e}$.

Whether or not the junction susceptance is negligible can be determined by running MICTEE. The value of B calculated by MICTEE should be compared to the admittances of all the transmission lines of the splitter. If B is not negligible, a compensation approach exists for equal power splitters where $Z_1 = Z_2$. A notch [9] is placed in the T-junction at the location indicated Figure 5.12. This reduces B enough to minimize its effect on the junction. The technique to determine the notch dimension, a, includes a numerical solution. No closed-form equations are available. The value of a varies with the line widths used in the junction. If the input line is wider than the output lines and for $\theta = 45$ degrees, then $a \approx 0.8W$ where W is the effective width (see (2.51)) of the output line. When the input line is equal to or narrower than the other lines, $a \approx 0.6W$. These are only guidelines and should be checked experimentally.

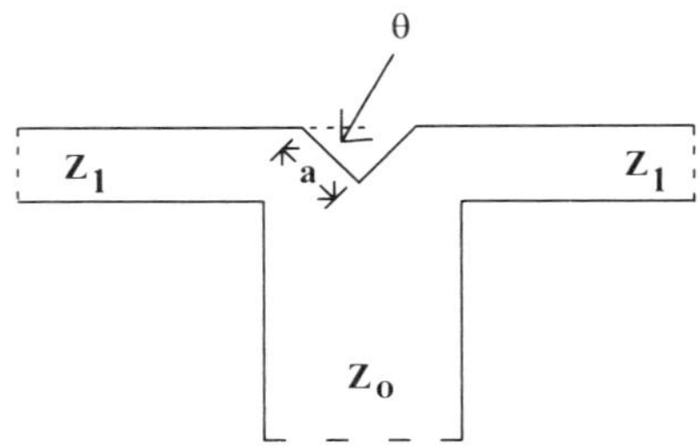

Figure 5.12 Notch compensation of an equal split T-junction.

5.8 CIRCULARLY POLARIZED DESIGN USING A SPLITTER

As with the hybrid-fed antenna, first determine the patch size and input impedance. Assume the same substrate and frequency as in Section 5.5. The patch is then a 6.302-cm square with an input impedance of 281Ω. Several options are available at this point. One is to transform the patch impedance to 50Ω. This uses the same transformer as in Section 5.5. The line width is 0.0898 cm, and its length is 3.624 cm. The T-junction must provide an equal power split and have output line impedances of 50Ω. In this situation, the splitter design is worked backward. With $Z_1 = Z_2 = 50\Omega$, an equal split will occur, and, ignoring the junction susceptance, the input impedance is 25Ω. The input is matched with a transformer with an impedance of $\sqrt{(25 \cdot 50)} = 35.4\Omega$. The required line width is 0.806 cm, while the length is 3.424 cm.

In addition to the splitter, a 90-degree-long section of line is needed to supply the phase shift. To ensure that the splitter operates correctly, the line should have a characteristic impedance of 50Ω. This results in both feeds presenting a 50-Ω impedance to the splitter. For a 50-Ω line, the width is 0.495 cm. A quarter-wavelength is 3.477 cm. The final step is to layout the lines. Figure 5.13 shows the layout. The quarterwave transformers are initially brought straight out from the patch. They are then bent 90 degrees. A chamfer (not shown) should be included. The change in the electrical length of the corner due to the chamfer should be accounted for. The program MMITER calculates this change. To avoid coupling between the patch and the feed lines, make the straight section out from the patch as long as possible and consistent with fitting in the rest of the network.

After the transformer come 50-Ω sections of line. Any bends in these lines should also be chamfered and the change in electrical length taken into account. The 50-Ω line for one section (in this case the feed on the right) must be a quarter-wavelength longer. The two 50-Ω lines meet at the junction. The junction input line is another quarterwave transformer. From there on, a 50-Ω line can be used to connect to the rest of the circuit. Note that the various line lengths are measured from the reference planes of the T-junction. Determine the reference plane locations using MICTEE. The positions of the reference

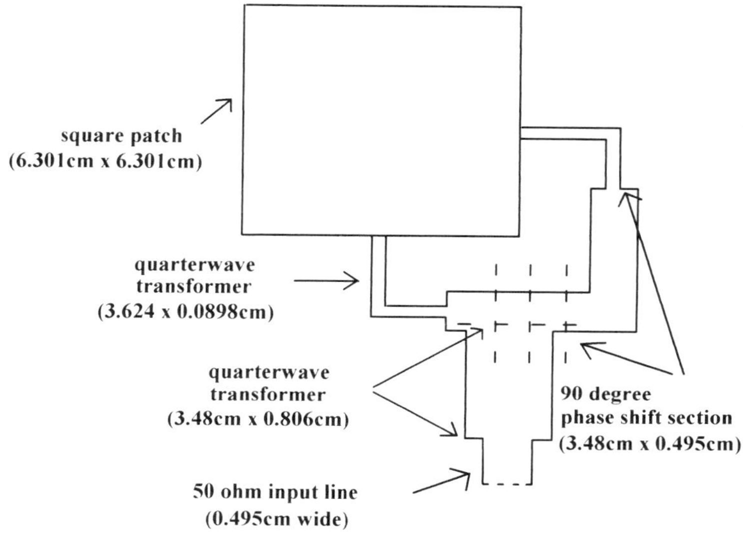

Figure 5.13 Design of splitter-fed circularly polarized patch.

planes will move if the T-junction is notched. In that case, the line lengths may have to be determined experimentally.

Another design uses transformers to convert the patch impedance to 100Ω instead of 50Ω. This is a line that is 0.0274-cm wide and 3.678-cm long with a characteristic impedance of 167Ω. The T-junction consists of a 50-Ω input line and 100-Ω output lines. The junction input impedance is therefore 50Ω. Lines of 100Ω (0.138-cm wide) connect the splitter to the patch transformers. The length of the extra 90-degree section is 3.597 cm. There would not be the need for a transformer at the input. This design is somewhat simpler and occupies less space provided the 167-Ω lines can be etched accurately enough. The bandwidth of splitter fed antennas is about 2% to 3%.

5.9 PROBE-FED PATCH DESIGN

The patches can also be fed with probes from underneath the ground plane as in Figure 3.1(b). Unlike microstrip feeds, probe feeds can be inset since they represent less of a disruption to the fields under the patch. This allows the freedom to select the patch impedance through choice of feed location. It is still a good idea not to use extreme insets as coupling between feeds could result. The coupling may change the feed amplitude and phase relationships and degrade the circular polarization. For the patch of the previous section, an inset of 1.829 cm produces a 100-Ω impedance. To obtain a 50-Ω value, the inset is 2.250 cm. The former is acceptable, but the latter may put the feeds too close together.

With probe feeds, the remainder of the feed network may be placed on another circuit board. This second board could be mounted directly to the antenna substrate, sharing a ground plane in a back-to-back fashion. If the probes are positioned to produce a 100-Ω impedance, a 70.7-Ω quarterwave transformer (3.536-cm long and 0.281-cm wide) will convert the patch impedance to 50Ω. Once a 50-Ω impedance level is reached, then either the hybrid of Figure 5.8 or the splitter of Figure 5.13 may be used to obtain the equal-amplitude, phase quadrature feed signals. Figure 5.14 shows an example using a splitter. It would also be possible to route 100-Ω lines from the antenna to an impedance-matching hybrid. Unlike the case presented in Section 5.5, the output side shunt line, which also has to be 100Ω, is easily realizable. Finally, the 100-Ω lines could be used with the splitter discussed at the end of the last section. This is probably the simplest design of all.

The interface region between the probe and the microstrip line on the feed substrate must be designed with some care. It is quite easy for this region to introduce parasitic reactances that cause impedance mismatches, signal loss, and other undesirable effects. In all cases, the combined thickness of the two substrates should be significantly less than a wavelength (for example, $\leq 0.05\lambda$). If the frequency is in the low microwave region or lower, it may be possible to simply butt the probe against the end of the line and solder

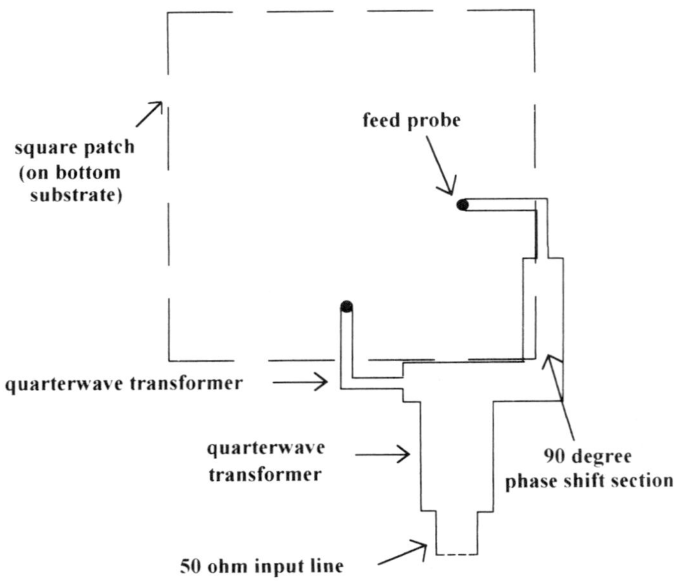

Figure 5.14 Probe-fed circularly polarized patch with splitter.

them together provided the probe diameter and the relief hole in the ground plane are smaller than the feed line width. An example is shown in Figure 5.15(a).

At higher frequencies, the line should be wider than the relief hole and extended beyond the probe by a quarter-wavelength (including the end-effect extension). See Figure 5.15(b). For the transition to work correctly, an electric field that is radially oriented between the probe and outside edge of the relief hole must be excited. The field must be uniform in magnitude across the aperture. Making the line wider than the hole plus completely covering the hole help create the desired field distribution. The quarterwave extension places a short circuit at the aperture, which also aids in establishing the correct fields. In addition, it provides a mechanism for fine tuning the transition.

The relief hole and probe diameters should be chosen so that the characteristic impedance of the coaxial structure they form is the same as that of the feed line. The ratio of the diameters is found from

$$Z_0 = \frac{60}{\sqrt{\epsilon_r}} \ln\left(\frac{r_h}{r_p}\right) \tag{5.31}$$

where Z_0 is the feed line characteristic impedance, ϵ_r is the substrate relative dielectric constant, r_h is the relief hole diameter in meters, and r_p is the probe diameter in meters. For very narrow line widths, this transition design becomes difficult to implement. The probe diameter may end up too small to be practical. In this case, the feed line can be widened in the vicinity of the transition as shown in Figure 5.16. The various dimensions of this design (W_t, L_1, and L_2) must be determined either by an electromagnetic analysis (for example, finite element programs like MSC/EMAS from MacNeal-Schwendler or HFSS from Hewlett-Packard) or by experiment. In all cases, it is good practice to verify any design experimentally.

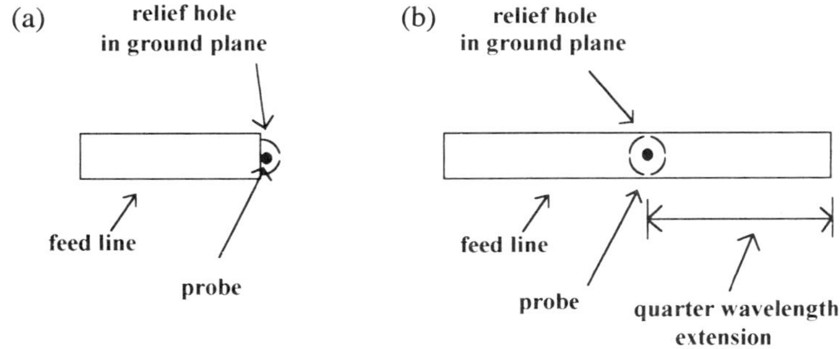

Figure 5.15 Low-frequency probe-to-feed line transitions: (a) basic transition and (b) improved transition.

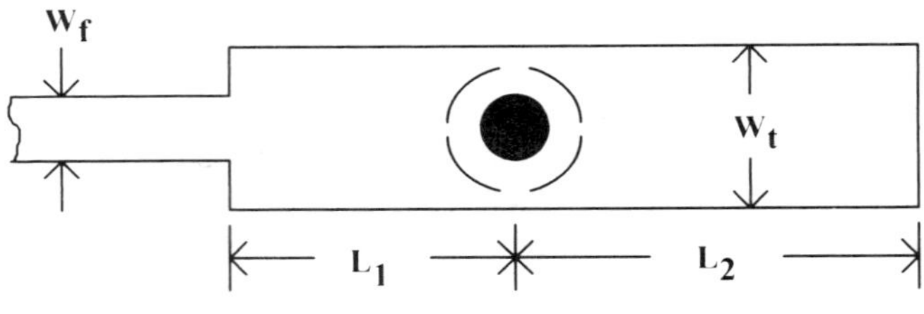

Figure 5.16 Modified transition for very narrow feed line widths.

References

[1] James, J. R., and P. S. Hall, eds., *Handbook of Microstrip Antennas*, London, UK: Peter Peregrinus, Ltd., 1989, Chap. 4.

[2] Sharma, P. C., and K. C. Gupta, "Analysis and Optimized Design of Single Feed Circularly Polarized Microstrip Antennas," *IEEE Trans. on Antennas and Propagation*, Vol. 31, No. 6, Nov. 1983, pp. 949–955.

[3] Richards, W. F., Y. T. Lo, and D. Harrison, "An Improved Theory for Microstrip Antennas and Applications," *IEEE Trans. on Antennas and Propagation*, Vol. 29, No. 1, Jan. 1981, pp. 38–46.

[4] Chang, K., ed., *Handbook of Microwave and Optical Components*, Vol. 1, New York, NY: John Wiley & Sons, 1989, pp. 145–150.

[5] Ho, C. Y., "Transform Impedance with a Branch-Line Coupler," *Microwaves*, Vol. 15, No. 5, May 1976, pp. 47–52.

[6] Gupta, R. K., and W. J. Getsinger, "Quasi-Lumped Element 3- and 4-Port Networks for MIC and MMIC Applications," *IEEE Int. Microwave Theory and Techniques Symp.*, 1984, pp. 409–411.

[7] Ho, C. Y., "Design of Lumped Quadrature Couplers," *Microwave J.*, Vol. 22, No. 9, Sept. 1979, pp. 66–70.

[8] Pozar, D. M, *Microwave Engineering*, Reading, MA: Addison-Wesley, 1990, pp. 391–393.

[9] Chadha, R., and K. C. Gupta, "Compensation of Discontinuities in Planar Transmission Lines," *IEEE Trans. on Microwave Theory and Techniques*, Vol. 30, No. 12, Dec. 1982, pp. 2151–2156.

Chapter 6
Broad Bandwidth Elements

The major limitation of microstrip antennas is narrow bandwidth. From the examples of Chapter 3, typical patch bandwidths are of the order of a few percent. Almost always, the factor that limits the bandwidth is the impedance behavior of the antenna. The frequency variation resembles that of a parallel resonant RLC circuit with a Q of around 60. Much effort has gone into broadening the impedance bandwidth. Some has involved attempting to realize the maximum bandwidth inherent in the antenna through better matching techniques. Other approaches broaden the bandwidth by using multiple elements that are either directly or parasitically coupled to the original patch. This chapter discusses the reasons for the narrow impedance bandwidth, some matching techniques, and several bandwidth-enhancing techniques using multiple elements. Several programs to assist in the design of broader bandwidth antennas are also presented.

6.1 LISTING OF COMPUTER PROGRAMS

The use of various techniques for increasing the bandwidth of a microstrip patch will be covered in this chapter. Several programs are introduced to assist in the design of a broader bandwidth patch using these techniques. The remaining programs mentioned in the chapter have been previously discussed. The programs introduced here are as follows.

CAPMAT: Determines the capacitance needed to help tune out the feed probe inductance of a patch built on an electrically thick substrate. Uses expressions given in [1] for the capacitance of two types of structures, namely, a disk capacitor and an annular slot capacitor. Calculates the dimensions of the capacitor. Also calculates the impedance versus frequency for a capacitor—matched patch using previously determined patch impedance data from a program such as PATCH9 and TLPATCH.

TLPATCH: Designs rectangular patches with two different dielectrics underneath the patch. Basically a modified version of PATCH9. Uses a transmission line model. The characteristic impedance, relative dielectric constant, and other parameters for the two-layer structure are determined from the quasi-static analysis given in [2].

MFPATCH: This program simply combines the impedance of two stacked patches in series to simulate the behavior of a stacked array. The impedance for the top patch is calculated using PATCH9 and then stored in a TOUCHSTONE compatible program. The lower patch impedance is found using PATCHC and again stored in a file. MFPATCH reads both files and combines the impedances. It is designed to provide the impedance of a stacked array where the resonant frequencies and patch sizes are sufficiently different such that no external mutual coupling occurs.

PARAPAT: Designs either a quarter-wavelength patch with one parasitic element or a half-wavelength patch with two parasitic elements for a wider impedance bandwidth. The parasitic element(s) are coupled via the radiating edge(s) of the driven patch. Calculates the required driven and parasitic patch lengths and the impedance versus frequency. Uses the theory of [3].

6.2 MICROSTRIP ANTENNA BANDWIDTH LIMITATIONS

The relatively rapid change in impedance with frequency is the major factor determining the patch bandwidth. Patches are resonant structures that look like a parallel RLC circuit in and around the resonant frequency. Like with any resonant circuit, the impedance behavior with frequency is closely linked to the patch Q. Typical Q's are around 60 to 70. Patches are one of many types of low-gain antennas; that is, antennas that have relatively broad beamwidths. These are antennas whose largest physical dimension is less than a half-wavelength or so. It can be shown [4] that for low-gain antennas, the antenna Q is inversely proportional to the volume occupied by the antenna as measured in terms of wavelengths. This is derived from Maxwell's equations and is a direct consequence of the laws obeyed by electromagnetic fields. There is therefore a minimum Q for any low-gain antenna that gets quite large as the antenna size shrinks. The only way to lower the Q is by adding loss (other than radiation). This is usually undesirable because introducing loss reduces efficiency, which is a concern when transmitting, and increases noise figure when receiving.

Since the bandwidth is determined by the patch physical size, one way to increase bandwidth is to increase the antenna size [5]. This approach is followed in bandwidth enhancement techniques that use multiple elements. Some of these will be discussed later in this chapter. For a single element, the size can be increased by making the antenna longer, wider, or thicker. The patch length is fixed by the requirement to be resonant. It can be lengthened only by lowering the substrate dielectric constant, which is usually desirable because of reduced surface-wave excitation. Increasing the patch width is possible as long as it remains less than a wavelength in the dielectric. Beyond a wavelength width, higher order modes may be excited that create pattern distortions and unpredictable impedance behavior. On a relative basis, widening the patch can produce fairly large bandwidth changes. As seen in Table 3.2, going from a nearly square patch to one that is a wavelength wide increases the bandwidth from 1.02% to 1.51% for an SWR requirement of 2.0:1 or less. From an absolute basis, the result is still a narrow bandwidth.

Perhaps the easiest way to increase the inherent patch bandwidth is by using a thick substrate. Table 6.1 shows bandwidth versus thickness for a substrate with an ϵ_r of 2.2. The bandwidth is for an SWR $\leq$ 2.0:1. The frequency is again 1.575 GHz. Notice that the bandwidth is almost a linear function of substrate height for these electrically thin substrates. The recommended maximum height as given by (3.1) is $h/\lambda_0 = 0.0324$. This appears to be the value where the surface wave is about 10% of the space wave and is a reasonable compromise between bandwidth and efficiency in most cases. If lower efficiencies can be tolerated, thicker substrates may be used for even wider bandwidths. The potential for coupling to other parts of the circuit and pattern degradation caused by the surface wave should be checked when thicker substrates are employed.

6.3 SIMPLE MATCHING APPROACHES

According to the laws of electromagnetics, the Q of a patch antenna is essentially fixed by its electrical size. Given this situation, it is natural to ask what is the maximum attainable bandwidth. Reference [6] presents a good discussion, part of which is covered here, of patch bandwidth and the capabilities of several matching approaches. Usually the bandwidth is defined as the frequency range over which the antenna SWR is less than or equal to some value. Let that value be S. Assume that the SWR $\leq S$ from the frequency

Table 6.1
Patch Bandwidth and Efficiency Versus Substrate Thickness

h/λ_0	*Bandwidth (%)*	*Radiation Efficiency (%)*	*Total Efficiency (%)*
0.010	1.54	96.3	88.0
0.020	3.01	92.9	89.8
0.030	4.60	89.7	87.9
0.040	6.28	86.8	85.6
0.050	8.08	84.1	83.2
0.060	10.11	81.6	81.0

f_l to f_u (where $f_l < f_u$). Also let the patch resonant frequency be denoted by f_r. The bandwidth is then given by

$$BW = \frac{f_u - f_l}{f_r} \tag{6.1}$$

At resonance, the patch input impedance is real. Let its value be R_0. In [6] it is shown that when connected to a transmission line with characteristic impedance, Z_0, the bandwidth can be expressed as

$$BW = \frac{1}{Q}\sqrt{\frac{(TS - 1)(S - T)}{S}} \tag{6.2}$$

where $T = R_0/Z_0$. Note the dependence on patch Q and the desired SWR.

For the antennas discussed in the previous chapters, two techniques for matching were sometimes mentioned. One is for antennas fed with a microstrip line and uses a quarterwave transformer connected to the patch edge. If the matched impedance is Z_0 and the patch resistance is R_0, the required transformer impedance can be found from $\sqrt{(R_0 Z_0}$. The transformer line length is $\lambda_0/(4\sqrt{\epsilon_{re}})$ where ϵ_{re} is the effective dielectric constant of the line. The other technique applies to probe-fed patches. The probe is inset from the patch edge to produce an impedance equal to that of the feed line. A microstrip feed can also be inset. For these cases, (6.2) reduces to (3.10)

$$BW = \frac{1}{Q}\frac{S - 1}{\sqrt{S}} \tag{6.3}$$

This is the equation used in program PATCHD to calculate patch bandwidth.

Both of the above matching approaches involve essentially a single-element matching network. It turns out that neither produces the maximum bandwidth for this type of elementary matching. The maximum bandwidth occurs when [6]

$$T = \frac{1}{2}\left(S + \frac{1}{S}\right) \tag{6.4}$$

To realize this value of T, the patch is matched to the impedance, R_0/T. This, of course, introduces a mismatch to the feed line whose impedance is Z_0. When the impedance-matching schemes of the previous paragraph are used, the patch becomes perfectly matched at one frequency, namely its resonant frequency. The SWR is thus very good in and around the resonant frequency but rapidly degrades as the frequency moves away from resonance. The mismatch introduced by (6.4) means the match is not as good at resonance but is better for frequencies removed from resonance. It's as though there is a certain amount of matching that can occur. The previous techniques place all of it at the resonant frequency to create a perfect match. Equation (6.4) "spreads" a poorer but still acceptable match over a band of frequencies.

The bandwidth obtained when (6.4) is satisfied is given by

$$BW = \frac{1}{2Q}\frac{\sqrt{S^4 - 1}}{S} \tag{6.5}$$

When the maximum SWR is 2.0:1, the bandwidth becomes $\approx 1/Q$.

The impedance-matching function of the quarterwave transformer may also be realized with a lumped-element network. This is useful at low frequencies where the length of distributed circuits can become prohibitive. Either T- or π-networks as shown in Figure 6.1 can be utilized. The elements are all reactive except for the terminating resistances. R_0 is the desired matching impedance and R_p is the patch resonant resistance. Equations for each element of the T-network are [7]

$$X_1 = \pm R_0\sqrt{\left(\frac{R}{R_0} - 1\right)} \tag{6.6}$$

$$X_2 = -\frac{R}{X_1/R_0 + X_3/R_p} \tag{6.7}$$

and

$$X_3 = \pm R_p\sqrt{\left(\frac{R}{R_p} - 1\right)} \tag{6.8}$$

where R is a parameter that controls the bandwidth of the circuit. R must be greater than the larger of R_0 or R_p. It is usually chosen to be about 5% above the larger quantity.

A similar set of equations exists for the π-network. Simply replace X_i with B_i, R_0 with $G_0 = 1/R_0$, R_p with $G_p = 1/R_p$, and R with G. G is again a bandwidth parameter that must be greater than the larger of G_0 or G_p.

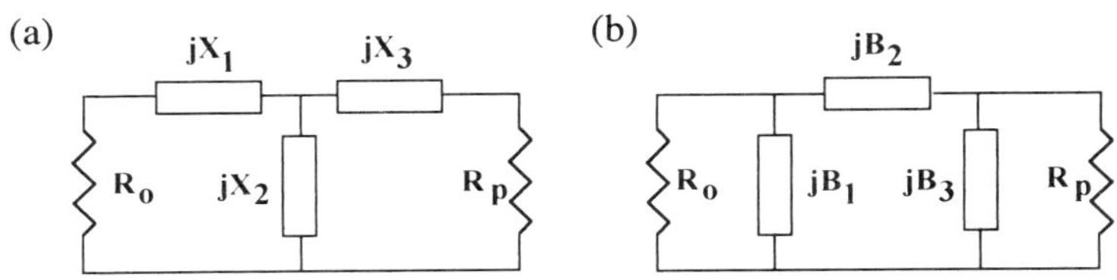

Figure 6.1 T- and π-networks of reactive elements for impedance matching: (a) T-network matching circuit and (b) π-network matching circuit.

6.4 OPTIMUM BANDWIDTH MATCHING CIRCUITS

It seems reasonable to expect that wider bandwidths could be obtained using more sophisticated matching approaches. This is indeed the case, but it is important to know that there are limits to the attainable bandwidth. Just as Maxwell's equations show that an antenna has a minimum Q based on its electrical size, the bandwidth over which a match can be obtained has limits as well [8]. These are again a result of the laws of circuit theory. For a given impedance, it is not possible to realize a perfect match over an arbitrary bandwidth. What can be realized is a constant (nonideal) match within a frequency range with an accompanying total mismatch outside this band. In [6], it is shown that for a parallel or series RLC circuit, the maximum attainable bandwidth for a given SWR is

$$BW_m = \frac{1}{Q} \frac{\pi}{\ln[(S + 1)/(S - 1)]} \tag{6.9}$$

Again note the inevitable presence of the Q of the impedance to be matched. To get an idea as to the improvement in bandwidth that is theoretically possible, the bandwidths of (6.3) and (6.9) are compared. Figure 6.2 shows the ratio of (6.9) to (6.3) as a function of the desired match, S [6]. The typical improvement is on the order of a factor of 4. This then represents the maximum attainable bandwidth for the patch using an optimum

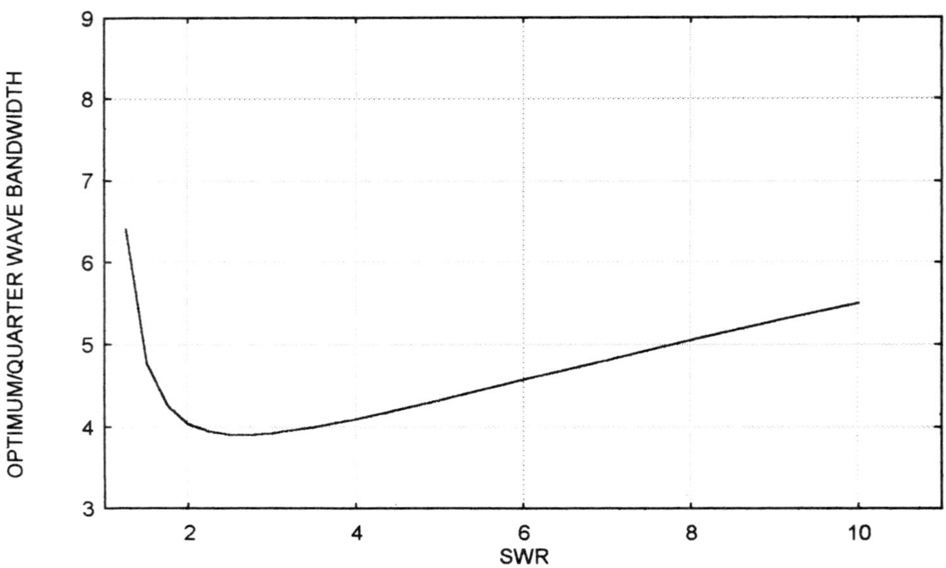

Figure 6.2 Ratio of optimum network to quarterwave transformer matching bandwidths.

matching circuit. The dramatic improvements indicated for very low values of S are somewhat misleading. The improvement increase is primarily due to a rapid decrease in the quarterwave match bandwidth. Overall the bandwidth is still quite small when S is small.

The optimum matching network is a filter that "absorbs" the reactive part of the patch impedance into the filter structure. Much of the theory and many useful design tables are presented in [9]. In addition, [6] gives an excellent discussion on the application of this technique to patch antennas. The design begins by determining the matching bandwidth and order of the filter. Although the degree of match and bandwidth improves with increasing filter elements, typically filters with 2 to 4 sections are used. Larger-sized filters do not offer much performance improvement and can be difficult to realize. The normalized low-pass elements for the prototype LC-ladder filter are found next, usually using published tables [9]. The low-pass structure is then transformed into the required band-pass filter.

There are many ways of realizing the filter. At low frequencies, lumped elements may be used. For higher frequencies, transmission lines serve as filter elements. The network designed in [6] consists of half-wavelength-long sections. The results in [6] show that the theoretical bandwidth improvements of Figure 6.2 can be approached. The practical realization of the matching circuit requires significant care for optimum results. When the filter is constructed using microstrip transmission lines, the various discontinuities must be properly accounted for; otherwise significant trial-and-error tuning is required.

6.5 CAPACITIVELY FED MICROSTRIP PATCHES

Increasing the substrate height is an easy and effective way of improving patch bandwidth. To keep surface-wave excitation to a minimum, the lowest possible dielectric constant material should be used. Patches on thick substrates are typically probe fed. To maintain a constant impedance, microstrip lines must be made wider as the substrate height increases. The line width can become too large and will affect the patch operation. Also feed line radiation rises and may interfere with the patch pattern. Using a probe feed eliminates these problems. Unfortunately the inductance added by the probe can push the input impedance completely into the inductive portion of the Smith chart. There no longer is a well-defined resonance frequency (in terms of impedance), and the impedance becomes difficult to match.

One way to compensate for the increased probe inductance is to add a capacitor in series. This can be done either externally in the feed network connected to the patch [10] or within the patch itself [1,11]. Two techniques have been proposed for the latter. The capacitor can be formed within the patch itself by etching a rectangular or circular gap around the feed probe as in Figure 6.3. The other is to form the capacitor by shortening the probe so that it no longer touches the patch and by adding a small disk to the end of the probe. The disk is separated from the patch by a dielectric layer that can have a

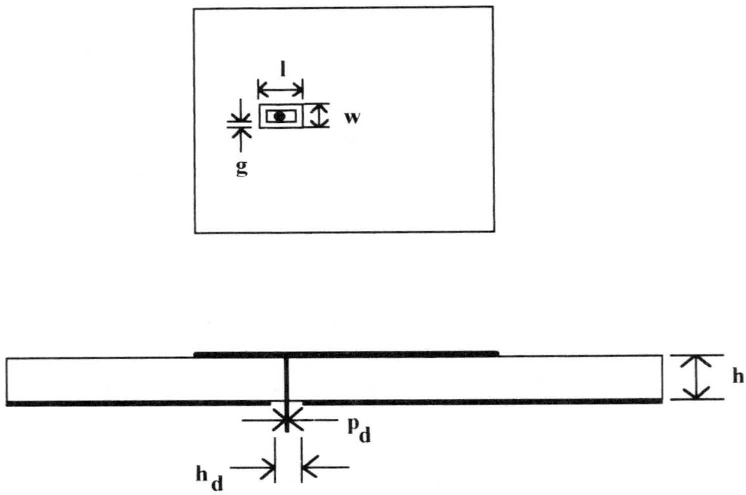

Figure 6.3 Rectangular patch with etched capacitor.

dielectric constant different from that of the original substrate. This is shown in Figure 6.4.

The etched capacitor is more easily implemented, but predicting the capacitance is more difficult. The disk capacitor is a more complicated structure, but its capacitance is calculated more easily.

The idea behind the capacitive feed is quite simple. At resonance, the inductance and capacitance inherent to the antenna equivalent circuit cancel each other out, leaving a real impedance. The capacitor is chosen such that its reactance is sufficient to cancel the residual reactance of the probe inductance. The required capacitance is

$$C_{\mathrm{m}} = \frac{1}{\omega_{\mathrm{r}}^2 L_{\mathrm{p}}} \tag{6.10}$$

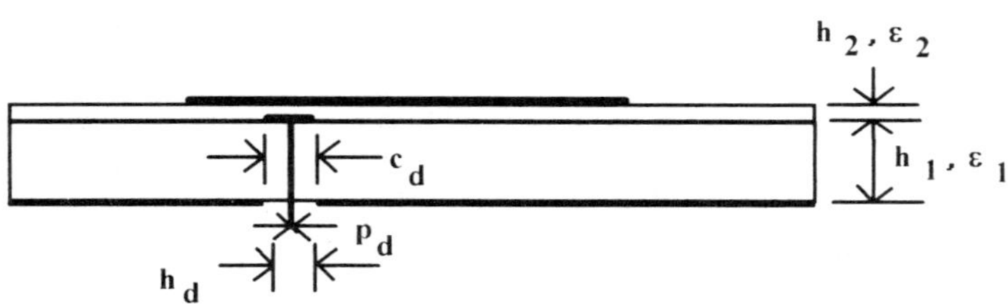

Figure 6.4 Patch fed with a disk capacitor.

where ω_r is the resonant frequency and L_p is the probe inductance. The extra capacitance thus brings the impedance back to resonance, and the wider bandwidth of the thicker substrate can be realized.

The probe inductance is given by the sum of (3.14) and (3.16). Reference [1] gives approximate expressions for the etched and disk capacitors. Let $a = c_d/2$ (see Figure 6.4) be the radius of the disk capacitor. The capacitance is

$$C_m = (a^2 \pi \epsilon_{r2} \epsilon_0 / h_2) \cdot \left\{ 1 + \left(\frac{2h_2}{\pi \epsilon_{r2} a} \right) \left[\ln\left(\frac{a}{2h_2} \right) + 1.41 \epsilon_{r2} + 1.77 + \left(\frac{h_2}{a} \right) (0.268 \epsilon_{r2} + 1.65) \right] \right\} \quad (6.11)$$

This expression is valid for $h_2/a < 0.5$. In (6.11), h_2 is the thickness of the dielectric between the disk and the patch and ϵ_{r2} is its relative dielectric constant. For the etched capacitor

$$C_m = \pi^{0.5} \left(\frac{w'}{l} \right) \epsilon_0 (\epsilon_r + 1)(w' + l) \quad (6.12)$$

where ϵ_r is the substrate relative dielectric constant; $w' = w - 2g$, with g being the gap width; and l and w are the capacitor length and width, respectively.

Equations (6.10) to (6.12) along with the probe model ((3.13) to (3.16)) of Chapter 3 have been programmed to provide a means of determining the value of the matching capacitor. The program is called CAPMAT. CAPMAT will be applied to broadband a 1.575-GHz rectangular patch. The first task is to determine a substrate thickness. PATCHD can be used to trade-off bandwidth and efficiency versus thickness. Assume a 10% bandwidth is desired. From Table 6.1, a substrate thickness of $0.060\lambda_0$ will have a 10% bandwidth. A review of the design requirements indicates that the 81% efficiency is acceptable and that no surface interaction problems are anticipated.

To design the thick antenna, the basic patch dimensions are determined from PATCH9 by not including the probe. If an attempt is made to design the patch and the feed probe is included, the program will not be able to find a resonant length because of the added probe inductance. There is too much inductance to cancel out by varying the length. As part of the procedure, the probe inset is varied until the desired input resistance at resonance is obtained.

A different program for determining the patch dimensions may be required for the disk capacitor approach. If the dielectric layer used for the capacitor has a different relative dielectric constant than the patch substrate, the patch dimensions and impedance can be calculated using program TLPATCH, which is a transmission line model that includes two dielectric layers under the patch. It operates in an identical manner as PATCH9 except for additional inputs for the second dielectric layer. Since it is identical, the steps involved in using TLPATCH will not be covered. When designing a patch with a thick substrate and a disk-matching capacitor, do not include the probe as with the previous example.

CAPMAT is then used to find the matching capacitor based upon the probe dimensions. CAPMAT works on an iterative basis. A set of dimensions and substrate parameters for the capacitor are entered along with the probe data. The probe inductance is calculated and from that the required value of capacitance determined. The capacitance for the input dimensions is also calculated. All three results are displayed. If the calculated result is not close enough to the required value, the user is given the option of re-entering data for a new capacitance calculation. After any calculation, the user can also quit. Results are displayed on the screen and options presented for printing and storing the data.

As an example, assume that a 10% bandwidth patch at 1.575 GHz is needed. The etched capacitor approach will be used. Based upon the data in Table 6.1, the substrate is chosen to have the parameters $\epsilon_r = 2.2$, tan $\delta = .001$, $h = 1.143$ cm ($0.060\lambda_0$), and t (line thickness) = 0.00356 cm. To obtain the patch dimensions, the *none* option is used for the feed type when running PATCH9. The patch width is selected to be 9.677 cm. Let the desired input impedance be 100Ω. This requires a feed inset of 0.490 cm. This results in a patch whose length is 5.331 cm with a resonant impedance of about 97Ω. The probe diameter is taken to be 0.127 cm with a relief hole diameter of 0.411 cm. These dimensions are for an SMA-type connector.

Running CAPMAT, the screen appears as follows.

CAPMAT.V10 8/28/95 13:20:10
[Current Program Date - 22/08/95]

Matching Capacitor for Probe Fed Patches

DO DISK (d) OR ETCHED (e) CAPACITOR? **e**

INPUT LENGTH AND WIDTH (cm) OF CAPACITOR PATCH? **1.524,1.016**

INPUT WIDTH OF ETCHED CAPACITOR GAP (cm)? **.0127**

INPUT PATCH SUBSTRATE THICKNESS (cm)? **1.143**

INPUT RELATIVE er AND loss tangent OF PATCH SUBSTRATE? **2.2,.001**

INPUT ANTENNA FREQUENCY (Ghz)? **1.575**

(At this point, the substrate relative dielectric constant is checked to see if it is within the limits, $1 \leq \epsilon_r \leq 10$. As before, if ϵ_r falls outside these limits, the user is given the option to change the value or continue. A similar check is made on the substrate height.)

INPUT FEED PROBE DIAMETER (cm)? **0.127**

INPUT FEED PROBE RELIEF HOLE DIAMETER (cm)? **0.411**

The equivalent circuit elements (see Figure 3.3) for the probe are calculated. The sum of the inductances, L_0 and L_p determines the probe inductance. The antenna frequency and inductance produce the required capacitance when inserted in (6.10). Next the program prints the probe inductance and required capacitance

PROBE INDUCTANCE = 8.247 nH
REQUIRED CAPACITANCE = 1.238 pF

and then evaluates (6.12) with the given inputs for the capacitor.

REQUIRED CAPACITANCE = 1.238 pF
CALCULATED CAPACITANCE = 0.821 pF

TRY ANOTHER CAPACITOR (y or n)? **y**

(Since the calculated value does not equal the required one, another iteration with different capacitor dimensions will be tried to get the correct capacitance.)

INPUT LENGTH AND WIDTH (cm) OF CAPACITOR PATCH? **1.765,1.407**

INPUT WIDTH OF ETCHED CAPACITOR GAP (cm)? **.0127**

A new calculation is performed, and the results again displayed.

PROBE INDUCTANCE = 8.247 nH

REQUIRED CAPACITANCE = 1.238 pF
CALCULATED CAPACITANCE = 1.237 pF

This time the correct value is obtained, so to exit the calculations answer no to the next question.

TRY ANOTHER CAPACITOR (y or n)? **n**

The capacitor and probe inductance can be combined with the patch impedance calculated by another program like PATCH9. For this example, PATCH9 was run without a feed (but with an inset) to obtain the patch dimensions. If the results are stored in a TOUCHSTONE compatible file, the patch impedance can be read by CAPMAT and combined with the capacitor and probe. This option is addressed with the following input.

CALCULATE PATCH IMPEDANCE WITH CAPACITOR (y or n)? **n**

For the time being, this option will be bypassed. At this point, the program displays the following results on the screen.

CAPMAT.V10 08-27-1995 21:38:48

Patch Matching Capacitor

Etched Capacitor

PATCH SUBSTRATE HEIGHT = 1.143 (cm)

PATCH SUBSTRATE er = 2.20 AND loss tangent = 0.0010

FEED PROBE DIAMETER = 0.127 (cm)

FEED PROBE RELIEF HOLE DIAMETER = 0.411 (cm)

PATCH FREQUENCY = 1.5750 (GHz)

CAP LENGTH = 1.765 (cm) AND WIDTH = 1.407 (cm)

CAPACITOR GAP = 0.0127 (cm)

The next inputs give the user the option to print out and store the results just as with previous programs like PATCH9. These will not be repeated here. The final input gives the option to do another case or quit the program.

The biggest factor affecting the etched capacitor value is its width. To increase the capacitance, increase the width. The length has a secondary effect, with a shorter value having a larger capacitance. The capacitance is virtually independent of the gap width.

If a disk capacitor is used, the program operates in an identical manner except for the inputs related to the capacitor. They are as follows.

INPUT RELATIVE er AND loss tangent OF CAPACITOR DIELECTRIC? **2.2,.001**

INPUT DIELECTRIC THICKNESS BETWEEN PROBE DISK AND PATCH (cm)? **.16**

INPUT DISK DIAMETER (cm)
[Diameter/Dielectric Thickness < 0.5]
.1016

(The disk capacitor model is valid for disk diameters that are less than half the thickness of the dielectric between it and the patch. If this ratio is exceeded, the user can re-enter the data if desired.)

The program proceeds in the same manner as with the etched capacitor. With the disk, the capacitance increases with increasing disk diameter and dielectric constant. Reducing the dielectric thickness also increases the capacitance.

Although the simple model describing the action of the matching capacitor provides significant insight into the operation of the antenna, it does not account for all behavior. The capacitance not only cancels the probe inductance, it also changes the input resistance. This is not predicted when the patch impedance, probe inductance, and capacitor are combined. Unfortunately, no simple explanation exists for the effect of the capacitor on the resistance. With thick substrates, radiation from the probe becomes a factor. The capacitor apparently modifies the probe current and radiation to cause the change in resistance. This modification occurs for most etched capacitors. For disk capacitors, the effect is most pronounced when the dielectric thickness is less than about 10% of the patch substrate. The input resistance is almost linearly proportional to the capacitor size provided it is less than about 10% of the patch length. The smaller the capacitor, the smaller the resistance.

It was mentioned earlier that CAPMAT can read in patch impedances from a file created by PATCH9 or TLPATCH and combine them with the probe and matching capacitor. This is done by answering yes when the program asks the following.

CALCULATE PATCH IMPEDANCE WITH CAPACITOR (y or n)? **y**

The program then requests the name of the file containing the impedance via

INPUT NAME OF TOUCHSTONE COMPATIBLE FILE (no extension)? **patch**

(CAPMAT assumes the file extension is .S1P).

Finally, the number of frequencies in the file is needed and requested by the following query.

INPUT NUMBER OF FREQUENCIES IN PATCH.S1P? **20**

When this option is selected, the output printout is displayed as before, listing, for example, the capacitor dimensions. The user is then requested to hit the enter key to see a listing of the impedances. At each frequency, the patch impedance with and without the capacitor is displayed. This is followed by the usual choices to print and store data.

Figure 6.5 shows the result of combining the previously determined capacitor with the patch. Two impedance loci are shown. The curve totally in the inductive portion of the Smith chart is that for the patch and the probe. The lower curve includes the capacitor. Not surprisingly, the latter represents simply the cancellation of the probe inductance. Because the resistance is also modified, the actual curve would lie further toward the left of the chart.

At this point, an experimental model should be built and the input impedance measured. Assuming the matching capacitor value is close to correct, the impedance locus should cross or be near to the real axis of the Smith chart. It will probably be at a lower resistance value than predicted because of the effect of the capacitor. The probe position should then be adjusted to obtain the correct input impedance. The amount of correction can be estimated using the measured results and simulations using PATCH9. If the impedance curve has not dropped down enough or has overshot into the capacitive portion of the Smith chart, the capacitor is either too small or too large. It can be experimentally modified to find the correct value with help from CAPMAT.

In spite of this inability to predict the actual resistance, CAPMAT is still useful for design purposes. It provides an initial estimate for the value of capacitance needed and approximate dimensions for its realization. In designing the antenna, the layer thickness for the disk approach should be kept as thin as possible (preferably < 10% of the patch substrate). For both capacitor types, use the smallest area capacitor consistent with the requirement to cancel the probe inductance. The capacitor is thus used to tune out the probe. For a given capacitor, the probe inset is then adjusted to produce the desired input resistance at resonance.

Large probe insets should be avoided. When the inset is near the center of the patch, the combination of the probe and patch acts more like a top-loaded monopole rather than

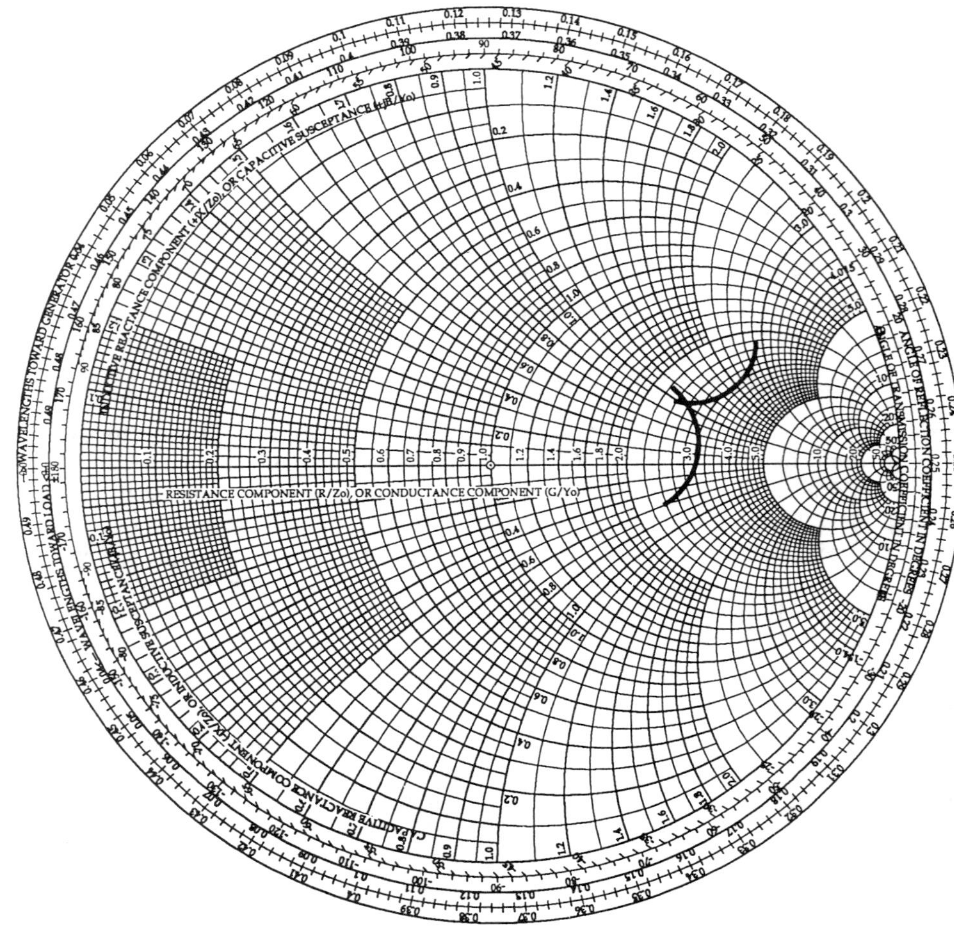

Figure 6.5 Input impedance versus frequency for the capacitively fed patch of this section.

a patch. The radiation patterns tend to look somewhat like a monopole over a ground plane with a partial null on-axis and power directed horizontally outward. Even for probe positions away from the patch center, some pattern distortions will occur primarily in the form of asymmetries, especially in the *H*-plane. Also significant cross-polarized radiation may occur. Both of these effects are caused by the thick substrate and long probe.

6.6 STACKED PATCH CONFIGURATIONS

Many techniques using multiple patches have been investigated to broaden the bandwidth. This includes multiple-patch configurations with all patches on the same substrate or stacked on top of each other. The former will be discussed in the next section. From the discussion of Section 6.1, the bandwidth broadening occurs because the size of the antenna increases. Most stacked patch configurations consist of two patches, one on top of the other. Three feed approaches have been tried with the stacked patches [5]. In one, the feed probe extends completely through the lower patch, passes through a hole in the lower patch, and connects to the upper patch. This configuration is shown in Figure 6.6.

The second type of feed only excites one of the patches, usually the lower one, as in Figure 6.7. In this configuration, the upper patch is parasitically coupled to the lower patch. The final configuration has two feeds—one for each patch, as in Figure 6.8. Each of these will be discussed in this section.

The stacked configuration of Figure 6.6 is particularly useful for applications requiring two frequencies spaced far enough apart that only one patch operates at a time. At frequencies well below or well above the resonant frequency, the impedance of a patch is very small and almost entirely reactive. The impedance locus essentially circles around the outside edge of the Smith chart. In Figure 6.6, the patches are in series with each other. At the resonant frequency of the lower patch, the upper patch looks basically like a short circuit. This short effectively connects the probe directly to the lower patch

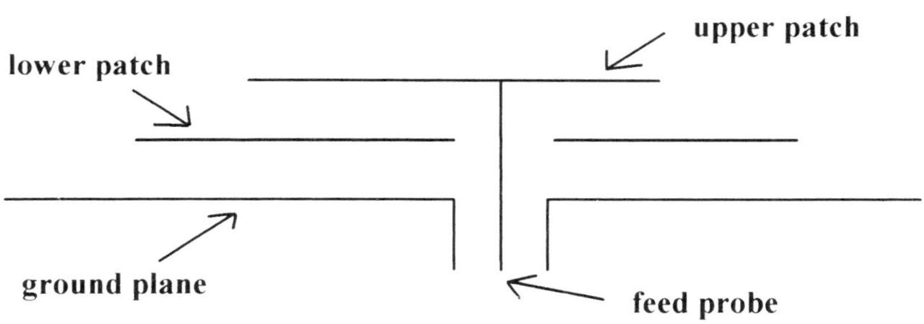

Figure 6.6 Stacked patches with common feed.

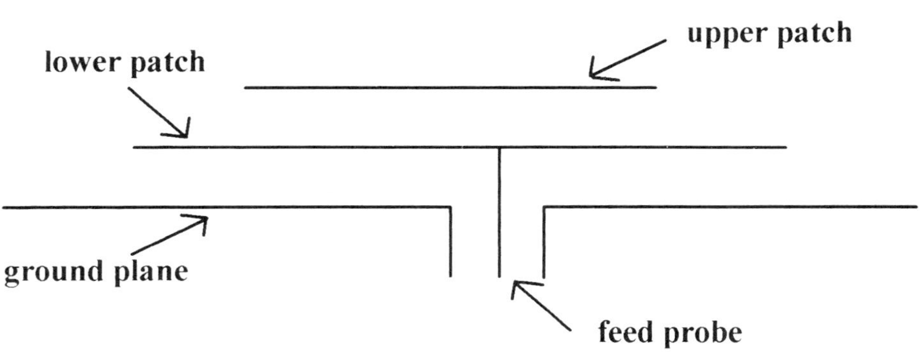

Figure 6.7 Parasitically stacked patches.

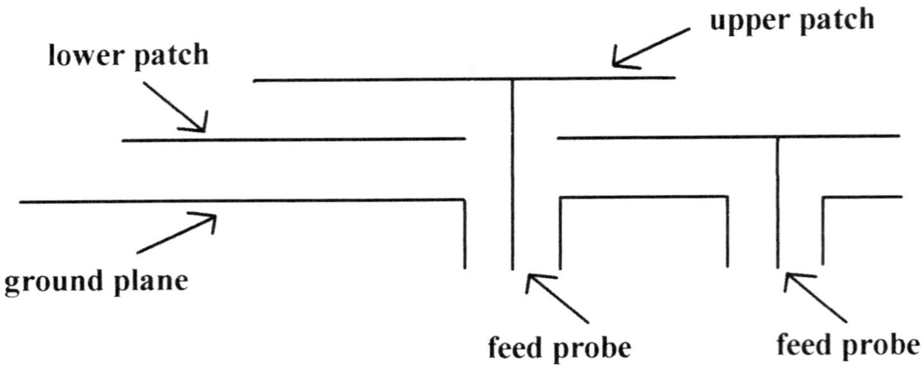

Figure 6.8 Stacked patches with individual feeds for each antenna.

metalization. Conversely at the upper patch resonant frequency, the lower patch appears as a short and thus connects the ground plane to the lower patch.

This simple model works reasonably well when the upper patch is the smaller of the two. In this situation there is little coupling between the patches via their external electromagnetic fields at the edges. It is important to keep the upper antenna smaller than the lower one; otherwise the antennas can become strongly coupled. The configuration of Figure 6.7 better exploits external coupling for improved bandwidth. The upper patch size can be minimized by using a higher dielectric constant substrate or using a quarterwave patch. When the upper patch is small enough, the lower patch appears like a ground plane to the upper antenna. The upper patch does not see the lower one until the patch sizes become comparable. At this point both the resonant frequency and resistance begin to decrease with increasing antenna size [5]. The lower antenna looks like a patch with a cover layer. Provided the upper patch is small, it operates as if other antenna were not present. As a rule of thumb, the length and width of the upper patch should be less than the corresponding lower antenna dimensions by two or more substrate thicknesses.

To illustrate the use of this approach, consider the design of a stacked patch to cover two frequency bands centered at 1.225 GHz and 1.575 GHz. These two frequencies correspond to those needed to receive both the codes for the GPS system. Assume the bandwidth needed for each band is that typically obtained from a single patch. The lower substrate is chosen to have a low dielectric constant to keep the size of the bottom patch large. Let the substrate have the parameters h = 0.160 cm, ϵ_r = 2.2, tan δ = 0.001, and t = .00356 cm. A higher dielectric constant substrate is used for the upper patch to reduce its size. Let the substrate's parameters be h = 0.160 cm, ϵ_r = 4.0, tan δ = 0.001, and t = .00356 cm. The feed probe has a diameter of 0.127 cm and a relief hole diameter of 0.411 cm. The lower patch is designed with PATCHC to include the effects of the top layer. For this example, the requirement for a circularly polarized patch for GPS will be ignored. A patch width of 12.032 cm is used. After several runs with PATCHC, it is found that a feed inset of 2.756 cm is needed to produce a 50-Ω impedance. The resonant length is 8.026 cm. The upper patch is designed with PATCH9. The width is taken to be 7.000 cm (same width-to-length ratio, 1.5, as the lower patch). The resonant length turns out to be 4.671 cm, and an inset of 1.524 cm produces a 50-Ω impedance. Figure 6.9 shows the resulting antenna configuration.

Since the antennas share a common feed, it is of interest to know what the impedance-versus-frequency curve of the combined structure looks like. The program MFPATCH is a simple program that reads in impedance-versus-frequency data from two TOUCHSTONE compatible files as generated by PATCHC and PATCH9 and then combines them in series. Both files must contain data at the same frequencies and have the same number of frequencies. In this design example, the impedance for both patches was calculated from 1.11 GHz to 1.67 GHz, and the results stored in files. MFPATCH is then run to see the combined impedance. The screen appears as follows.

MFPATCH.V10 08/31/95 10:14:57
[Current Program Date - 21/08/95]

Calculates impedance vs. frequency for two stacked patches sharing a common feed and forming a multi-frequency antenna

NOTE
REQUIRES A TOUCHSTONE COMPATIBLE FILE FOR EACH PATCH WITH THE SAME FREQUENCY SPAN. FILES MUST HAVE FOUR HEADER LINES

(The program assumes four header lines before the impedance versus frequency data. Any program described in this book that is capable of generating TOUCHSTONE files will be compatible with this requirement.)

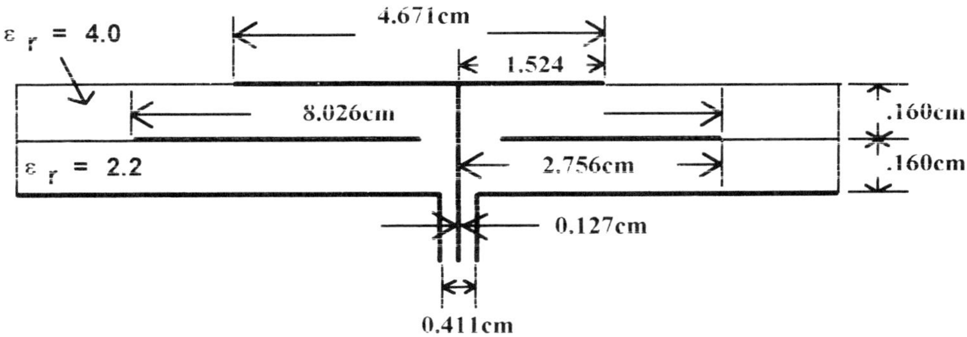

Figure 6.9 Stacked element array for operation at 1.225 Ghz and 1.575 Ghz (upper patch is 7.000-cm wide and lower patch is 12.032-cm wide).

INPUT NUMBER OF FREQUENCIES TO BE ANALYZED? **225**

INPUT NAME (no extension) OF FILE CONTAINING UPPER PATCH DATA?
lower

(The files will always have the extension .S1P.)

INPUT NAME (no extension) OF FILE CONTAINING LOWER PATCH DATA?
upper

The program opens the files, reads in the data, combines them in series, then outputs the results.

MFPATCH.V10 11-05-1995 21:01:40
Current Program Date - 21/08/95

Multi-frequency patch input impedance vs. frequency

DATA FILE FOR UPPER PATCH - c: data winword upper.S1P
DATA FILE FOR LOWER PATCH - c: data winword lower.S1P

FREQUENCY (GHz)	*INPUT IMPEDANCE (Ohms)*	*FREQUENCY (GHz)*	*INPUT IMPEDANCE (Ohms)*
1.110	3.670 J 10.460	1.392	4.620 J 9.430
1.112	3.690 J 10.570	1.395	4.640 J 9.530
1.115	3.700 J 10.680	1.397	4.650 J 9.620
1.117	3.720 J 10.780	1.400	4.650 J 9.720
1.120	3.740 J 10.900	1.402	4.660 J 9.810
1.122	3.760 J 11.010	1.405	4.680 J 9.910
1.125	3.770 J 11.130	1.407	4.690 J 10.000
1.127	3.790 J 11.250	1.410	4.690 J 10.090
1.130	3.820 J 11.380	1.412	4.690 J 10.270
1.132	3.840 J 11.510	1.415	4.710 J 10.370
1.135	3.870 J 11.660	1.417	4.720 J 10.470
1.137	3.890 J 11.800	1.420	4.740 J 10.570
1.140	3.920 J 11.960	1.422	4.750 J 10.670
1.142	3.950 J 12.120	1.425	4.770 J 10.760
1.145	3.980 J 12.280	1.427	4.770 J 10.870
1.147	4.010 J 12.460	1.430	4.790 J 10.980
1.150	4.040 J 12.640	1.432	4.800 J 11.080
1.152	4.080 J 12.840	1.435	4.820 J 11.190
1.155	4.130 J 13.060	1.437	4.840 J 11.290
1.157	4.180 J 13.270	1.440	4.860 J 11.400
1.160	4.230 J 13.510	1.442	4.870 J 11.520

1.162	4.280 J	13.750	1.445	4.890 J	11.630
1.165	4.340 J	14.030	1.447	4.900 J	11.750
1.167	4.410 J	14.320	1.450	4.920 J	11.880
1.170	4.490 J	14.620	1.452	4.950 J	12.000
1.172	4.580 J	14.960	1.455	4.970 J	12.130
1.175	4.680 J	15.330	1.457	4.990 J	12.260
1.177	4.790 J	15.720	1.460	5.020 J	12.390
1.180	4.930 J	16.160	1.462	5.040 J	12.530
1.183	5.090 J	16.640	1.465	5.070 J	12.670
1.185	5.270 J	17.160	1.467	5.100 J	12.820
1.188	5.480 J	17.750	1.470	5.130 J	12.970
1.190	5.750 J	18.420	1.472	5.160 J	13.130
1.192	6.070 J	19.150	1.475	5.180 J	13.290
1.195	6.470 J	20.000	1.477	5.220 J	13.460
1.197	6.980 J	20.960	1.480	5.260 J	13.640
1.200	7.620 J	22.070	1.482	5.300 J	13.830
1.202	8.460 J	23.350	1.485	5.340 J	14.020
1.205	9.590 J	24.840	1.487	5.390 J	14.210
1.207	11.140 J	26.570	1.490	5.440 J	14.420
1.210	13.330 J	28.560	1.492	5.490 J	14.650
1.212	16.510 J	30.740	1.495	5.550 J	14.880
1.215	21.240 J	32.850	1.497	5.620 J	15.120
1.217	28.280 J	34.100	1.500	5.690 J	15.380
1.220	38.110 J	32.430	1.502	5.760 J	15.650
1.222	48.910 J	24.390	1.505	5.850 J	15.930
1.225	54.150 J	9.040	1.507	5.940 J	16.240
1.227	48.930 J	–6.230	1.510	6.050 J	16.560
1.230	38.290 J	–14.220	1.512	6.160 J	16.910
1.232	28.640 J	–15.960	1.515	6.290 J	17.290
1.235	21.700 J	–14.830	1.517	6.430 J	17.680
1.237	17.010 J	–12.820	1.520	6.600 J	18.120
1.240	13.840 J	–10.720	1.522	6.790 J	18.580
1.242	11.650 J	–8.800	1.525	7.010 J	19.090
1.245	10.090 J	–7.120	1.527	7.260 J	19.630
1.247	8.960 J	–5.670	1.530	7.560 J	20.23
1.250	8.110 J	–4.420	1.532	7.900 J	20.880
1.252	7.470 J	–3.330	1.535	8.310 J	21.610
1.255	6.960 J	–2.390	1.537	8.810 J	22.400
1.257	6.560 J	–1.560	1.540	9.400 J	23.280
1.260	6.240 J	–0.820	1.542	10.120 J	24.250
1.262	5.980 J	–0.170	1.545	11.020 J	25.320
1.265	5.760 J	0.410	1.547	12.150 J	26.510

1.267	5.590 J	0.950
1.270	5.440 J	1.420
1.272	5.310 J	1.860
1.275	5.200 J	2.250
1.277	5.110 J	2.610
1.280	5.030 J	2.960
1.282	4.960 J	3.270
1.285	4.900 J	3.560
1.287	4.850 J	3.830
1.290	4.810 J	4.090
1.292	4.760 J	4.330
1.295	4.730 J	4.560
1.297	4.700 J	4.770
1.300	4.680 J	4.970
1.302	4.650 J	5.170
1.305	4.630 J	5.350
1.307	4.620 J	5.530
1.310	4.600 J	5.700
1.312	4.580 J	5.860
1.315	4.570 J	6.020
1.317	4.560 J	6.170
1.320	4.550 J	6.310
1.322	4.550 J	6.450
1.325	4.540 J	6.590
1.327	4.540 J	6.720
1.330	4.540 J	6.840
1.332	4.530 J	6.970
1.335	4.530 J	7.090
1.337	4.530 J	7.210
1.340	4.530 J	7.330
1.342	4.520 J	7.450
1.345	4.520 J	7.560
1.347	4.530 J	7.660
1.350	4.530 J	7.770
1.352	4.530 J	7.870
1.355	4.530 J	7.980
1.357	4.540 J	8.080
1.360	4.540 J	8.190
1.362	4.550 J	8.280
1.365	4.550 J	8.380
1.367	4.550 J	8.480
1.370	4.560 J	8.570

1.550	13.570 J	27.810
1.552	15.410 J	29.200
1.555	17.800 J	30.640
1.557	20.940 J	32.010
1.560	25.060 J	33.070
1.562	30.380 J	33.310
1.565	36.920 J	31.940
1.567	44.050 J	27.820
1.570	50.060 J	20.180
1.572	52.500 J	9.780
1.575	50.140 J	–0.590
1.577	44.270 J	–8.210
1.580	37.300 J	–12.360
1.582	30.910 J	–13.820
1.585	25.680 J	–13.690
1.587	21.610 J	–12.760
1.590	18.500 J	–11.490
1.592	16.110 J	–10.140
1.595	14.280 J	–8.820
1.597	12.840 J	–7.590
1.600	11.710 J	–6.450
1.602	10.800 J	–5.420
1.605	10.070 J	–4.490
1.607	9.470 J	–3.640
1.610	8.970 J	–2.880
1.612	8.560 J	–2.180
1.615	8.210 J	–1.550
1.617	7.910 J	–0.960
1.620	7.660 J	–0.430
1.622	7.440 J	0.060
1.625	7.250 J	0.510
1.627	7.080 J	0.930
1.630	6.940 J	1.310
1.632	6.810 J	1.670
1.635	6.700 J	2.010
1.637	6.600 J	2.330
1.640	6.510 J	2.620
1.642	6.430 J	2.900
1.645	6.360 J	3.160
1.647	6.300 J	3.410
1.650	6.240 J	3.640
1.652	6.190 J	3.860

1.372	4.570 J	8.670	1.655	6.150 J	4.070
1.375	4.580 J	8.770	1.657	6.110 J	4.270
1.377	4.590 J	8.860	1.660	6.070 J	4.460
1.380	4.590 J	8.960	1.662	6.030 J	4.640
1.382	4.600 J	9.060	1.665	6.000 J	4.810
1.385	4.600 J	9.150	1.667	5.980 J	4.970
1.387	4.610 J	9.240	1.670	5.950 J	5.130
1.390	4.620 J	9.340			

PLOT OUT RESULTS ON SMITH CHART (y or n)? **y**

The Smith chart plot of the results is given in Figure 6.10. The impedance locus does a loop. Each time it crosses the real axis near the 50-Ω point represents the resonant frequency of one of the patches. As with other programs, the user can print out and store the output. Keep in mind that MFPATCH simply combines the impedances of the individual patches. The results will only be useful when there is enough separation between antenna sizes to minimize the effect of external coupling.

The configuration of Figure 6.7 is most useful when attempting to broaden the bandwidth of the upper patch. The patches form a pair of coupled resonators, and the resulting impedance locus also loops around like in Figure 6.10. This time the coupling via the external patch fields is exploited to broaden the bandwidth. By properly choosing the lower patch size, the resonant frequencies of the coupled system can be brought closer together. The loop becomes smaller, meaning less impedance variation with frequency. The probe position is then adjusted to bring the loop toward the center of the Smith chart. The lower patch size is usually slightly smaller or sometimes equal to the upper patch. This maximizes the bandwidth [5], because the coupling is a function of the relative patch sizes, and helps bring the system resonant frequencies closer. Bandwidth improvements of several percent are common.

Various other configurations of this type have been reported in the literature. If the space between the patches is filled with a relatively thick, low dielectric constant material, bandwidths of up to 20% can be achieved [12]. These antennas also show increased gain (narrower radiation patterns) because they form an array in the direction perpendicular to the ground plane. The ground plane creates images of the patches, so the array has four elements associated with it.

Unfortunately the parasitically stacked patches represent an extremely complicated electromagnetic problem that requires significant computational effort to solve. Usually these structures are designed experimentally. It is important when using larger spacings to keep the dielectric constants low to avoid surface waves. In many cases, foam is used with the patches printed on thin, low dielectric constant substrates.

The antenna of Figure 6.8 is a variant of Figure 6.6. The two separate feeds provide additional flexibility. Outputs are available from both antennas simultaneously. The feeds can be placed such that the antennas have the same or orthogonal polarizations [5]. The upper patch behaves essentially like an isolated antenna, provided it is reasonably smaller

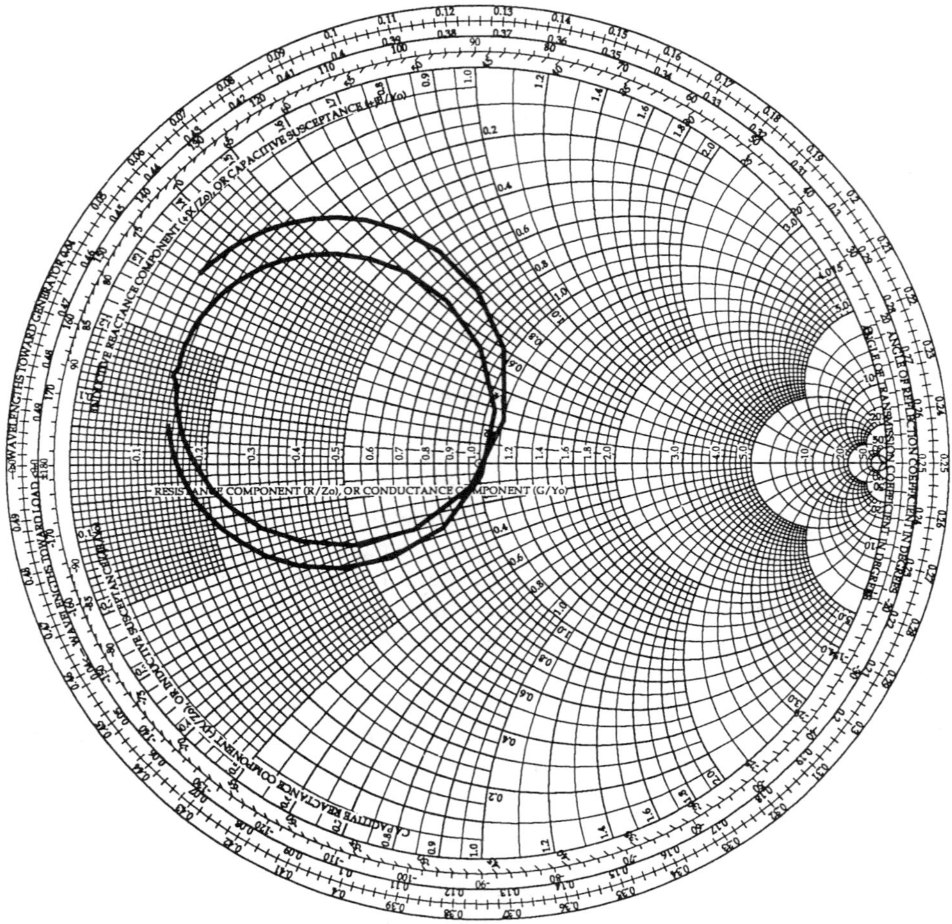

Figure 6.10 Impedance versus frequency for the stacked patch configuration of Figure 6.9.

than the lower patch. If possible, the upper feed should pass through the center of the lower patch. The electric field is zero here, so the feed has minimum impact on the lower patch. If the upper feed does not pass through the center, it acts like an inductive post inside the lower patch. The resonant frequency is shifted somewhat by the post. With a smaller upper antenna, the lower patch looks much like an isolated structure. The upper patch can be designed using PATCH9, while PATCHC is appropriate for the lower antenna. Again, a quarterwave element may be suitable for the upper antenna.

6.7 COPLANAR PATCH CONFIGURATIONS

Another way of increasing patch antenna bandwidth is by placing additional elements alongside the patch. As with the stacked patches, several configurations are possible. Elements can be placed along the patch sides, at the patch ends, and in both locations as shown in Figure 6.11. Coupling between the driven and parasitic elements occurs because of the fringing electromagnetic fields associated with the patches. The lengths of the parasitic elements are made slightly different than the driven patch and sometimes with respect to each other. They therefore have resonant frequencies that are slightly displaced from the driven element. This creates loops in the impedance versus frequency when seen on a Smith chart. These loops reduce the frequency variation, thus increasing the bandwidth. The size of the loops is a function of the spacings between the driven and parasitic elements. Gap spacings are typically 0.3 to 1.0 times the substrate height with narrower spacings used along the nonradiating edges. The location (on the Smith chart) of the loops depends on the feed location. Depending upon which configuration is adopted, bandwidth improvements of 3 to 6 times that of a single patch have been realized [13,14].

The additional elements may be directly coupled to the driven patch as in Figure 6.12 [15]. Microstrip lines are used to interconnect the patches. Since the electric field is zero at the patch center, the lines used to couple patches along the nonradiating edges must be positioned off-center. This configuration has more flexibility than the parasitic coupling of Figure 6.11. The length, width, and location of the feed lines provide more ways to control the coupling between elements. The line lengths are generally taken to be slightly larger than two substrate thicknesses. This is to ensure that the capacitive coupling due to the fringing fields is small enough to be ignored. The only coupling is thus via the lines. Again the outside elements have lengths different from the driven patch.

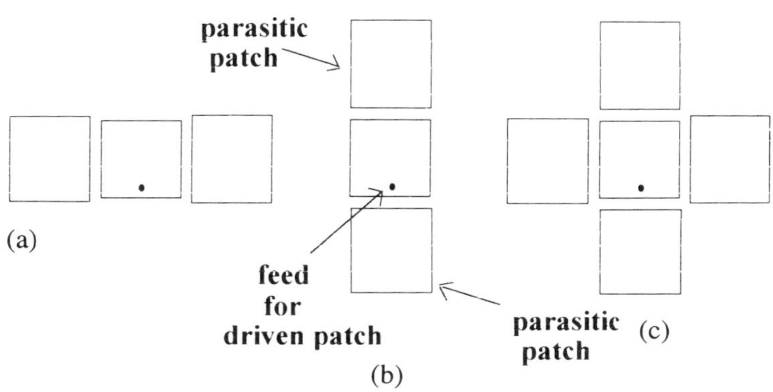

Figure 6.11 Parasitically coupled patches for increased bandwidth: (a) nonresonant side, (b) resonant side, and (c) all four sides.

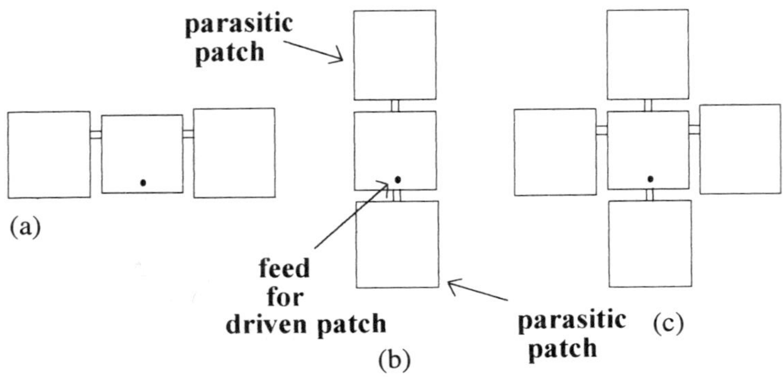

Figure 6.12 Direct coupled patches for increased bandwidth: (a) nonresonant side, (b) resonant side, and (c) all four sides.

Bandwidth improvements of 5 to 7 times that of an isolated patch have been realized [15].

Both of these approaches produce improved bandwidths by making the antenna larger as required by Maxwell's equations. It is the coupling between different frequency resonators that causes the bandwidth to increase. From a radiation standpoint, the *antenna* is now the array of elements. The relative amplitudes and phases of the radiation from each element vary greatly with frequency. This is due to the nature of the coupling, particularly for the parasitic configurations. As a result, the radiation patterns vary significantly over the frequency bandwidth where the impedance is reasonably matched. Nulls can occur in the pattern, and the pattern peak may not be broadside to the antenna. This is the major drawback to using these schemes for broadening the bandwidth. They improve the impedance bandwidth but degrade the pattern bandwidth. Also these are electromagnetically complicated structures that are not amenable to simple analysis. The primary design approach is to experimentally determine element sizes and spacings.

One version of Figure 6.11(a) has been reported [5] that provides a compromise between impedance bandwidth improvement and pattern degradation. The elements adjacent to the nonradiating sides of the patch are relatively narrow strips as seen in Figure 6.13. Less radiation occurs from the strips because of their narrow width; therefore, patch patterns are less effected. The Q's of the strips are much higher than the patch, so they exert less influence on the impedance behavior and thus smaller bandwidth enhancement is seen. Best results occur when the parasitic strips are slightly longer than the patch and with gap separations of 2.5 to 3 times the substrate thickness. Using different lengths for each strip broadens the bandwidth further. The operating frequency drops by a few percent when the strips are present. The SWR = 2.0:1 bandwidth increased from about 1% to 3.3%. Measured radiation patterns show little change in the E-plane and a small skewing in the H-plane. The asymmetry in the H-plane is fairly small, being only 2 dB or so in the worst case.

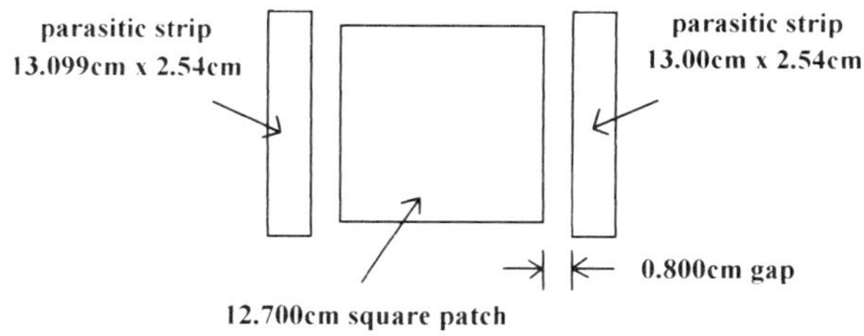

Figure 6.13 UHF parasitically coupled patch on a 0.318-cm-thick Teflon-fiberglass substrate [5].

A variant of Figure 6.11(b) has also been developed that, like the antenna of Figure 6.13, increases bandwidth without major pattern distortions. This concept uses quarterwave parasitics coupled to the radiating edges as seen in Figure 6.14 [3,16]. In Figure 6.14(a), each parasitic element is half the length of the driven patch. For Figure 6.14(b) both elements have identical lengths. The use of short-circuited elements eliminates one of the radiating edges of the parasitic element. Also the gap between elements is kept small, so effectively the edges of the patch and parasitic merge into one radiating slot. As before, the bandwidth improvement is not as great, on the order of a factor of 2. By the nature of their construction, these antennas are probe fed.

The slot between the patch and parasitic radiates and couples the elements. It still looks like a parallel combination of a conductance and capacitance. Instead of going to ground as in a normal patch, the slot equivalent circuit connects the patch and parasitic in series. All the radiation is represented in the slot equivalent circuit, so both the patch

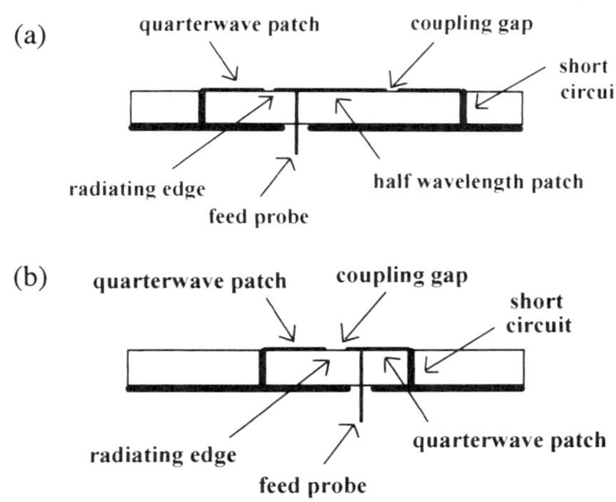

Figure 6.14 Radiating edge-coupled parasitic patches with low pattern distortion: (a) parasitically coupled half-wavelength patch and (b) parasitically coupled quarter-wavelength patch.

and parasitic simply appear as resonators. A fairly simple equivalent circuit can be established for the antenna [3]. Figure 6.15 shows one for the quarterwave patch. A similar circuit holds for the halfwave patch. The parallel LC circuits represent the patch and parasitic resonators.

Assuming that the feed is at the edge of the driven patch, it is possible to derive an expression [3] for the input admittance seen by the probe.

$$Y_e = \frac{GB^2 + j[B^3 + 3B_cB^2 + 2(G^2 + B_c^2)B}{G^2 + (B + B_c)^2} \tag{6.13}$$

where B is the equivalent susceptance of the LC circuits

$$B = j\omega C + \frac{1}{j\omega L}$$

but since both patches are in reality short circuited transmission lines

$$B = \frac{-j}{Z_0 \tan(2\pi\sqrt{\epsilon_{re}}l_p/\lambda_0)}$$

with Z_0 and ϵ_{re} being the characteristic impedance and effective dielectric constant of the patch transmission line and l_p the patch length. Also $B_c = \omega C_c$ is the susceptance of the slot capacitance.

In Figure 6.15, V_1 and V_2 are the voltages across the parasitic and patch, respectively. From the equivalent circuit, their ratio is [3]

$$\frac{V_2}{V_1} = \frac{G + jB_c}{G + j(B + B_c)} \tag{6.14}$$

The difference in the voltages is the voltage across the conductance and therefore is responsible for radiation from the antenna. Resonance is defined in the customary manner

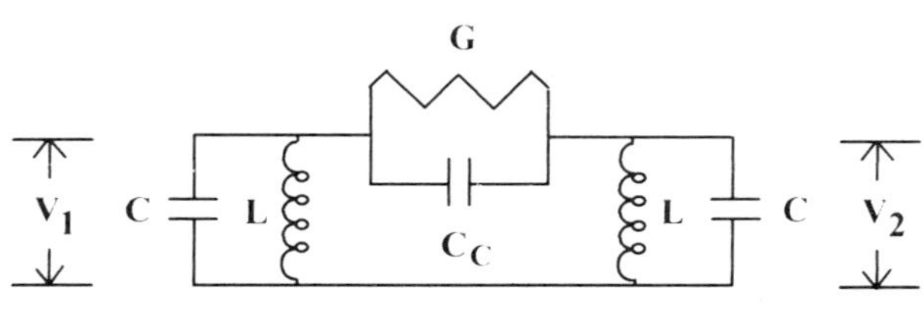

Figure 6.15 Equivalent circuit for parasitically coupled quarterwave patch.

by requiring the imaginary part of (6.13) to be zero. As with other patches, the patch length is adjusted to obtain resonance. Since the imaginary part of (6.13) is a cubic in B, there are three solutions, each yielding a patch length.

For one of the solutions, $V_2/V_1 = 1$, so there is no radiation. For another, $V_2 \gg V_1$, so essentially the parasitic is not excited and the antenna looks like a single patch. The third resonance occurs when [3]

$$B = \frac{-3B_c - \sqrt{B_c^2 - 8G^2}}{2} \tag{6.15}$$

and in this case $V_2/V_1 = -1$. This is equivalent to twice the voltage across the radiating slot of the patch when the parasitic is not present [16]. Four times the power is thus radiated. Each element stores an equal amount of energy, which is twice the value of the patch alone [16]. The resulting Q is half that of the isolated patch, resulting in a doubling of the bandwidth. Note that when (6.15) is satisfied, B is negative, which means that the antenna will be resonant at a lower frequency compared to the original patch. Also the impedance is one-fourth that of the single patch.

The radiation properties of the antenna should be virtually identical to the isolated patch. For the quarterwave case, radiation still occurs from a single slot. Its characteristics are very similar to the radiating edge of the patch. Similarly for the half-wavelength case, the two slots are at the same locations as they would be for the patch alone.

The design of these parasitically coupled antennas is aided by using program PARAPAT, which uses (6.13) to (6.15) to determine the required patch length and the input impedance as a function of frequency. To use the equivalent circuit model for design, expressions are needed for the slot conductance and capacitance. PARAPAT approximates the slot conductance using (2.29). This is a reasonable approximation since it is derived from a slot in an infinite ground plane.

The slot capacitance is more difficult to estimate. The usual expressions for a gap in a microstrip line [17] are only valid for width-to-substrate height ratios less than 2. At the desired resonance, the electric fields that go to ground on each side of the slot are out of phase. There are also fringing fields that span the gap between the elements. The field distribution is therefore not unlike that for the odd mode of a pair of coupled microstrip lines. The total capacitance of the odd mode contains a term that represents the capacitance of the fringing fields between the lines. It seems reasonable that this capacitance approximates that for the slot.

Reference [17] contains expressions for the fringing field capacitance. There are two components—one for the fringing fields in the air above the substrate, C_{ga}, and one for the fields in the dielectric, C_{gd}. The gap capacitance is the sum of the two. For C_{ga} [17]

$$C_{ga} = \begin{cases} \dfrac{\epsilon_0}{\pi}\ln\left[2\dfrac{1+\sqrt{k'}}{1-\sqrt{k'}}\right] & 0 \le k^2 \le 0.5 \\ \\ \dfrac{\pi\epsilon_0}{\ln\left[2\dfrac{1+\sqrt{k}}{1-\sqrt{k}}\right]} & 0.5 \le k^2 \le 1.0 \end{cases} \tag{6.16}$$

where $k = (s/h)/\{(s/h) + 2(l_p/h)\}$ and $k'^2 = 1 - k^2$. In the expression for k, s is the slot (gap) width between patch and parasitic, h is substrate height, and l_p is the patch (and parasitic) length.

The other component is given by [17]

$$C_{gd} = \frac{\epsilon_0\epsilon_r}{\pi}\ln\left[\coth\left(\frac{\pi s}{4h}\right)\right] + 0.65C_f\left[\frac{0.02}{s/h}\sqrt{\epsilon_r} + \left(1 - \frac{1}{\epsilon_r^2}\right)\right] \tag{6.17}$$

where $2C_f = \sqrt{\epsilon_{re}}/cZ_0 - \epsilon_0\epsilon_r l_p/h$ with Z_0 and ϵ_{re} being the characteristic impedance and effective dielectric constant of a microstrip line of width l_p. These expressions yield the capacitance per unit length. The capacitance of the slot is found by multiplying (6.16) and (6.17) by the patch width and then adding the two terms.

PARAPAT calculates the patch characteristic impedance and effective dielectric constant using (2.1) to (2.16) and then includes dispersion. The feed probe is handled using the model of Section 3.3. Finally, if pins are used to short circuit the parasitic elements, the length is adjusted according to (3.17).

PARAPAT will be applied to a case presented in [3]. When PARAPAT is run, the screen appears as follows.

PARAPAT.V16 02/09/95 08:56:34
[Current Program Date - 24/08/95]

Calculates Patch with Parasitic Elements for Wider Bandwidth

INPUT DESIGN FREQUENCY (GHz)? **1.29**

DESIGN QUARTER (q) OR HALF WAVELENGTH (h) PATCH? **h**

USE A PERFECT SHORT (s) OR SHORTING PINS (p)? **p**

The following inputs are only requested when the option **p** is selected.

INPUT SHORTING PIN DIAMETER (cm)? **.159**

INPUT CENTER-TO-CENTER SPACING BETWEEN PINS (cm)? **.750**

INPUT SUBSTRATE HEIGHT (cm)? **.318**

(PARAPAT does not have checks built into it regarding substrate thickness, for example, so it is up to the user to be aware that the model validity may be violated for certain inputs. As with all the models, PARAPAT is valid for thin substrates.)

INPUT SUBSTRATE RELATIVE er AND LOSS TANGENT? **2.5,.001**

INPUT CONDUCTOR THICKNESS (cm)? **.00356**

INPUT CONDUCTOR RELATIVE (to copper) CONDUCTIVITY
[Default = 1]
?<**enter**> (selects default value of 1)

INPUT GAP WIDTH (cm)? **.100**

INPUT PATCH WIDTH (cm)? **6.401**

INCLUDE FEED PROBE (y or n)? **n**

(If answered **y**, the program asks for probe and relief hole diameters.)

INPUT FEED POSITION (cm)
[0 <= lf <= 3.180]
[slot patch center]
?1.842

(PARAPAT calculates the patch length using the result of (6.15) before requesting this input. The effect of the inset is included as a cosine-squared function of the inset as in (3.12).)

INPUT START AND STOP FREQUENCIES FOR IMPEDANCE CALCULATION?
1.19,1.39

INPUT NUMBER OF FREQUENCIES TO CALCULATE IMPEDANCE? **20**

Equation (6.13) is evaluated at each requested frequency. The resulting output to the screen is as follows.

PARAPAT.V16 09-02-1995 09:15:01
Current Program Date - 24/08/95

HALFWAVE PATCH WITH TWO PARASITICS
Substrate height (cm) = 0.318 Line thickness (cm) = 0.00356
Relative dielectric constant = 2.500 Loss tangent = 0.0010
Patch width (cm) = 6.401 Patch length (cm) = 6.361
Parasitic element length (cm) = 3.180
Feed point inset (cm) = 1.842
Design frequency (GHz) = 1.290000

FREQUENCY (GHz)	*INPUT IMPEDANCE (ohms)*		
1.190	0.953	J	8.729
1.195	1.069	J	9.113
1.200	1.206	J	9.533
1.205	1.369	J	9.997
1.210	1.565	J	10.511
1.215	1.802	J	11.083
1.220	2.093	J	11.724
1.225	2.455	J	12.446
1.230	2.910	J	13.264
1.235	3.494	J	14.193
1.240	4.254	J	15.254
1.245	5.265	J	16.463
1.250	6.638	J	17.830
1.255	8.544	J	19.338
1.260	11.240	J	20.899
1.265	15.094	J	22.245
1.270	20.515	J	22.710
1.275	27.546	J	20.919
1.280	34.746	J	14.985
1.285	38.403	J	4.721
1.290	35.784	J	−5.910
1.295	29.097	J	−12.585
1.300	22.132	J	−15.019
1.305	16.593	J	−14.979
1.310	12.578	J	−13.889
1.315	9.729	J	−12.476
1.320	7.693	J	−11.048
1.325	6.211	J	−9.716
1.330	5.108	J	−8.512
1.335	4.271	J	−7.434
1.340	3.623	J	−6.468
1.345	3.112	J	−5.600
1.350	2.703	J	−4.813
1.355	2.370	J	−4.094
1.360	2.097	J	−3.432
1.365	1.869	J	−2.816
1.370	1.677	J	−2.237
1.375	1.514	J	−1.688
1.380	1.375	J	−1.162
1.385	1.255	J	−0.654
1.390	1.150	J	−0.157

These results agree well with the calculations in [3] using the inset value of 1.842 cm. No data was given in [3] about feed position or probe dimensions. Reference [3] used a similar expression for slot conductance but estimated the capacitance using experimental data. Since the impedance at resonance is directly related to the slot conductance, it must be assumed that some sort of inset was used to obtain the reported input impedance. The experimental results in [3] confirmed the doubling of the bandwidth when the parasitics were present. The options to plot impedance on a Smith chart, print out the results, and store the data are available.

The gap width affects the resistance at resonance since a characteristic of the model is that at resonance the impedance is approximately $1/(4G)$ for the quarterwave patch and $1/(8G)$ for a halfwave one. G is the conductance of the slot. As the slot width increases, the slot radiates more power and therefore G decreases. The input impedance becomes higher. From (6.15), the slot capacitance must be such that $B_c^2 \geq G^2$ for there to be a solution for B. The capacitance decreases with gap width. From the analysis in [3], best results occur when $B_c > 5G$. To ensure this condition, the gap width should be less than the substrate thickness. In the example given here, the gap is about one-third the thickness.

When designing for wider bandwidths, using a thicker substrate is probably the first option to consider. Assuming that the increased level of surface-wave excitation and somewhat reduced efficiency is acceptable, this is the easiest approach to implement. CAPMAT may be used to help compensate for the probe impedance. Keep in mind that surface waves will not only corrupt the pattern but also cause coupling between different parts of the circuit. The configurations of Figures 6.13 and 6.14 are good choices for situations in which the increased substrate height will not produce the desired bandwidth or when the increased surface wave is not acceptable. The stacked patch of Figure 6.7 is a good choice as well but suffers from a lack of design information. The implementation of this approach will probably require a fair amount of experimentation.

For applications with two frequency bands with a separation of at least the patch bandwidth, the approach of Figure 6.6 is useful. It is simple and straightforward to design. If increased flexibility is needed, then the dual feed configuration in Figure 6.8 may be tried, especially if the feed for the top patch passes through the center of the lower patch.

References

[1] Bernard, R., R. Tchanguiz, and A. Papiernik, "Capacitors Provide Input Matching of Microstrip Antennas," *Microwaves & RF*, Vol. 33, No. 7, July 1994, pp. 103–106.

[2] Svacina, J., "Analysis of Multilayer Microstrip Lines by a Conformal Mapping Method," *IEEE Trans. on Microwave Theory and Techniques*, Vol. 40, No. 4, April 1992, pp. 769–772; plus corrections, *IEEE Trans. on Microwave Theory and Techniques*, Vol. 40, No. 11, Nov. 1992, p. 2116.

[3] Wood, C., "Improved Bandwidth of Microstrip Antennas Using Parasitic Elements," *IEE Proc. Pt. H (Microwaves, Optics and Acoustics)*, Vol. 127, No. 4, Aug. 1980, pp. 231–234.

[4] Chu, L. J., "Physical Limitations of Omni-Directional Antennas," *J. Appl. Phys.*, Vol. 19, Dec. 1948, pp. 1163–1175.

[5] James, J. R., and P. S. Hall, *Handbook of Microstrip Antennas*, London, UK: Peter Peregrinus, Ltd., 1989, Chap. 6.

[6] Pues, H. P., and A. R. Van de Capelle, "An Impedance-Matching Technique for Increasing the Bandwidth of Microstrip Antennas," *IEEE Trans. on Antennas and Propagation*, Vol. 37, No. 11, Nov. 1989, pp. 1345–1354.

[7] Boulouard, A., "Lumped Element Phase Shifting and Matching Elements," *Rf Design*, Vol. 11, No. 7, July 1988, pp. 58–60.

[8] Fano, R. M., "Theoretical Limitations on the Broadband Matching of Arbitrary Impedances," J. Franklin Inst., Vol. 249, Nos. 1–2, Jan.–Feb. 1950, pp. 57–83 and 139–154.

[9] Matthaei, G. L., L. Young, and E. M. T. Jones, *Microwave Filters, Impedance—Matching Networks and Coupling Structures*, New York, NY: McGraw-Hill, 1964, pp. 120–135.

[10] Griffin, J. M., and J. R. Forrest, "Broadband Circular Disc Microstrip Antenna," *Electron. Lett.*, Vol. 18, No. 6, March 1982, pp. 266–269.

[11] Vandenbosch, G. A. E., and A. R. Van de Capelle, "Study of the Capacitively Fed Microstrip Antenna Element," *IEEE Trans. on Antennas and Propagation*, Vol. 42, No. 12, Dec. 1994, pp. 1648–1652.

[12] Chen, C. H., A. Tulintseff, and R. M. Sorbello, "Broadband Two-Layer Microstrip Antenna," *IEEE Antennas & Propagation Symp. Digest*, 1984, Boston, MA, pp. 251–254.

[13] Kumar, G., and K. C. Gupta, "Broadband Microstrip Antennas Using Additional Resonators Gap-Coupled to the Radiating Edges," *IEEE Trans. on Antennas and Propagation*, Vol. 32, No. 12, Dec. 1984, pp. 1375–1379.

[14] Kumar, G., and K. C. Gupta, "Nonradiating Edges and Four Edges Gap—Coupled Multiple Resonator Broad-Band Microstrip Antennas," *IEEE Trans. on Antennas and Propagation*, Vol. 33, No. 2, Feb. 1985, pp. 173–178.

[15] Kumar, G., and K. C. Gupta, "Directly Coupled Multiple Resonator Wide-Band Microstrip Antennas," *IEEE Trans. on Antennas and Propagation*, Vol. 33, No. 6, June 1985, pp. 588–593.

[16] James, J. R., P. S. Hall, and C. Wood, *Microstrip Antenna Theory and Design*, London, UK: Peter Peregrinus, 1981, pp. 106–107.

[17] Gupta, K. C., R. Garg, and I. J. Bahl, *Microstrip Lines and Slotlines*, Norwood, MA: Artech House, Inc., 1979, pp. 338–339.

Chapter 7
Microstrip Antenna Arrays

The previous chapters have dealt with various single-element microstrip antennas. The patterns from patch antennas are very broad. Some applications require narrow beamwidth antennas to provide increased range, rejection against interference, and isolation. The beamwidth of an antenna is a function of its size in terms of wavelengths. Microstrip antennas can be arrayed to produce narrow beamwidths. This chapter discusses the beamwidth and gain properties of arrays and commonly used architectures for array design. In addition, programs to compute array performance and assist in design will be demonstrated.

7.1 LISTING OF COMPUTER PROGRAMS

The designs of various types of microstrip arrays will be covered in this chapter. A number of programs are described that assist in the design of these arrays. Some are for any type of array, while others are for specific architectures. The following describes the programs to be presented in the chapter.

ARRAYCAL: Estimates the beamwidth and directivity of linear and planar two-dimensional arrays. Several amplitude distributions are available to the user to assess the impact of tapering. Standard array theory as presented in [1] and [2] is used for the calculations. The program is useful for performing an initial sizing of the array. Factors such as number of elements and element spacings can be found to meet beamwidth and directivity requirements.

APERDIST: Determines the amplitude distribution given the number of elements and array length. Provides the user with the choice of several amplitude distributions in order to trade-off sidelobe level and beamwidth. Calculates and displays the resulting pattern. Uses standard array expressions for the distribution and pattern calculations. Useful for finding the amplitude distribution needed to meet a sidelobe level and beamwidth requirement. Stores results in a file.

NARRAYD: A general-purpose pattern calculator for linear arrays. Handles virtually any array. Element spacings, amplitudes, and phases can be totally arbitrary. Array data can be entered either via the keyboard or a file. Calculates the radiation pattern for the array. An element pattern can be included in the calculations. Both rectangular and polar plots are available. For the former, several options relating to absisca and ordinate scales are available. The pattern may be stored in a file. Useful for investigating the effect of such factors as element spacing, number of elements, amplitude and phase distributions, and excitation errors.

RESFEED: Determines the element conductance for a series-fed resonant array when the individual radiators can be represented as a shunt conductance across the feed line. Applies (7.32), (7.35), and (7.36). Requires the number of elements in the array, the impedance of the feed line, and the desired amplitude distribution, which may be entered either by the keyboard or a file. Does either end or center fed arrays. Outputs both the normalized and unnormalized element conductances required to realize the distribution. Results may be placed in a file. This program is used as part of a series array design to convert from an aperture distribution (needed for a particular pattern) to an element conductance distribution.

PATCHCOM: A modified version of PATCH9 for series-fed array calculations. Uses the same transmission line model as PATCH9, but assumes a feed line at both radiating edges as would exist in a series fed array. Calculates the slot conductance, patch characteristic impedance, effective dielectric constant, and length over a range of user input patch widths. Useful for determining the patch sizes needed for each of the element conductances calculated by RESFEED or CONDCAL to realize a given aperture distribution. Running PATCHCOM is one of the last steps in a resonant array design. The output is used to determine the patch dimensions for each element. Results may be stored in a file.

CONDCAL: Calculates the element conductances needed to realize a given aperture distribution for a series-fed traveling array. As with RESFEED, it is valid for elements that can be represented as shunt conductances across the feed line. Uses (7.47) to find the conductances. Requires as inputs the number of elements, element spacing, feed line impedance, feed line attenuation (optional), power into the load, and excitations either from the keyboard or a file. Outputs both normalized and unnormalized conductances. Stores results in a file if desired. This program performs the same function as RESFEED except for a traveling wave array. It converts an amplitude distribution into a conductance distribution.

7.2 ARRAY THEORY

To derive some of the characteristics of arrays, consider two antennas positioned on the x-axis of a Cartesian coordinate system as shown in Figure 7.1. The antennas are separated by the distance, d. Assume that they are excited with equal amplitudes and in phase. For simplicity, the discussion will be restricted to the (x-z)-plane; therefore, only pattern variations with θ will be considered. Because Maxwell's equations are linear, the total field radiated by the two antennas at some point in the far field is given by the sum of the fields from each antenna. Let the total field be denoted by $E(\theta)$; then

$$E(\theta) = E_1(\theta)\frac{e^{-jkr}}{r} + E_2(\theta)\frac{e^{-jkr_1}}{r_1} \tag{7.1}$$

where $E_1(\theta)$ and $E_2(\theta)$ are the far-field patterns of each antenna and $k = 2\pi/\lambda_0$. r and r_1 are the distances from the antennas to the point in the far field. When the point is located at a sufficient distance away, r and r_1 are essentially parallel to each other as indicated in the figure.

Equation (7.1) holds for two arbitrary antennas. It also holds if the antennas are identical. Because of mutual coupling, the far-field pattern of one antenna is affected by the presence of the other and vice versa. If the environment seen by one antenna is not exactly the same as that seen by the other, then the individual antenna patterns will not be the same even if the antennas are. Fortunately in many instances the mutual coupling is small, so the patterns are identical for all practical purposes. In that case, $E_1(\theta) \approx E_2(\theta)$. The distances, r and r_2, are not the same, but when the observation point is far enough away, the difference in their magnitudes is very small. Since the denominators in (7.1) only affect the magnitude of the expression, it is reasonable to replace r_1 with r.

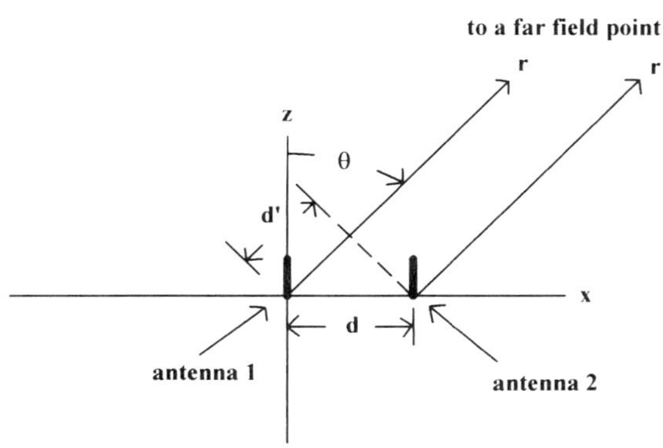

Figure 7.1 A two-element array and associated geometry.

In exponential terms, the small differences cannot be ignored. The phase changes by 2π for every wavelength difference. Typical path length differences are on the order of a half-wavelength. When r and r_1 are nearly parallel, the path length difference can be found from the geometry and is given by

$$d' = r - r_1 \approx d \sin(\theta) \tag{7.2}$$

the total field becomes

$$E(\theta) = E_1(\theta)\frac{e^{-jkr}}{r}[1 + e^{jkd \sin(\theta)}] \tag{7.3}$$

This expression consists of three terms multiplied together. The middle term is a consequence of the spatial spreading of the field with distance. There is no θ variation, so this term is not of concern here. The first term is the pattern of the element, and the last is due to the array. This multiplication of terms is an application of a concept called pattern multiplication, which is widely used to determine array patterns when the elements are identical and mutual coupling is negligible. The last term is called the array factor. In most cases, the element pattern is very broad. For large arrays, the pattern variation with θ is dominated by the array factor.

To increase the flexibility of the array, each element may be excited with a different amplitude and phase. Since phase is relative, let the element at the origin be the reference; that is, its phase is zero. With a_i and β_i being the amplitude and phase of the excitation of the ith element, (7.3) becomes

$$E(\theta) = E_1(\theta)\frac{e^{-jkr}}{r}[a_1 + a_2 e^{-j(\beta_2 - kd \sin(\theta))}] \tag{7.4}$$

Consider the case in which the elements are spaced a half-wavelength, $a_1 = a_2$, and $\beta_2 = 0$. The far-field pattern is given in Figure 7.2. At $\theta = 0$ degrees and $\theta = 180$ degrees, the path lengths, r and r_1, are exactly equal. With equal, in-phase excitation, the fields from the two antennas add to produce a pattern maximum. Conversely at $\theta = 90$ degrees, the difference in path lengths is 180 degrees (they are spaced a half-wavelength apart), and the fields cancel. A similar cancellation occurs at 270 degrees. When the same antennas are excited out of phase ($\beta_2 = -180$ degrees), the maxima are at 90 degrees and 270 degrees and the minima at 0 degrees and 180 degrees. This pattern is shown in Figure 7.3. As a final example of the pattern for a two-element array, Figure 7.4 gives the pattern when the elements are a quarter-wavelength apart with a –90-degree phase shift. This time the pattern assumes a cardioid shape and has only one maximum as seen in Figure 7.4.

Equation (7.4) can be extended to cover the general case with N elements. The far-field pattern is simply the sum over all elements

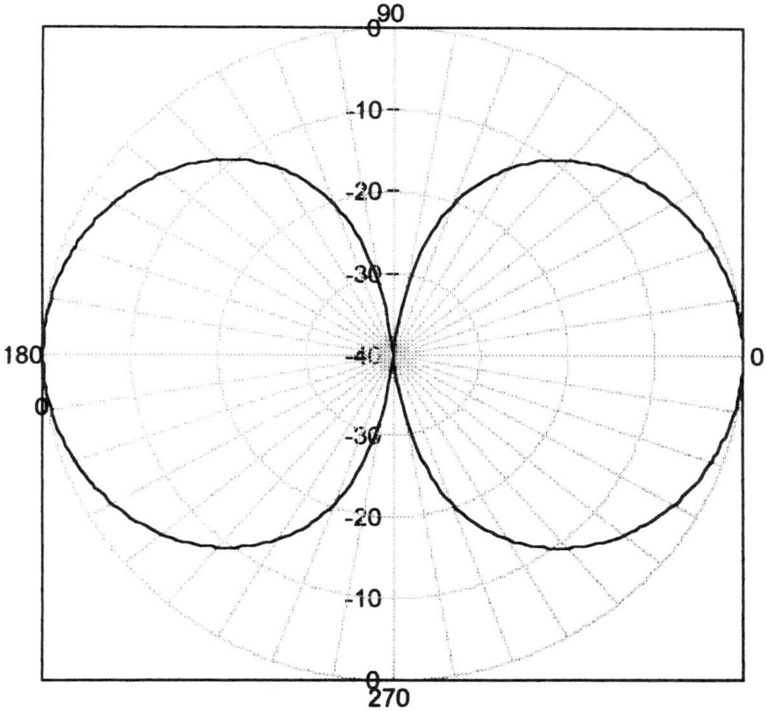

Figure 7.2 Two-element array pattern for in-phase, equal amplitude excitation (half-wavelength spacing).

$$E(\theta) = E_e(\theta)\sum_{n=1}^{N} a_n e^{j[(n-1)kd\sin(\theta)-\beta_n]}\frac{e^{-jkr}}{r} \tag{7.5}$$

In (7.5), $E_e(\theta)$ is the element pattern. For many cases, phase shift between elements is a constant, so $\beta_n = (n-1)\beta$. Then (7.5) becomes

$$E(\theta) = E_e(\theta)\sum_{n=1}^{N} a_n e^{j(n-1)[kd\sin(\theta)-\beta]}\frac{e^{-jkr}}{r} \tag{7.6}$$

This is thus the pattern for an N-element array with equal element spacing and a constant phase shift between elements as determined by the method of pattern multiplication. Most of the arrays dealt with in practice are adequately modeled using (7.6).

To study some of the properties of an array, let each element be excited equally in amplitude. After some algebraic manipulations, (7.6) reduces to

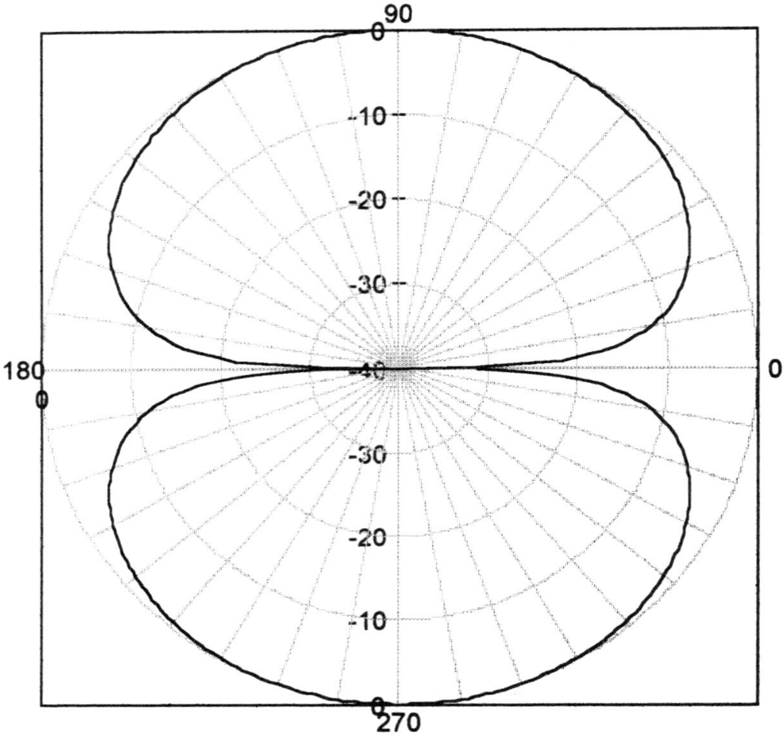

Figure 7.3 Two-element pattern for out-of-phase, equal amplitude excitation (half-wavelength spacing).

$$E(\theta) = \frac{1}{N} \frac{\sin[N(kd \sin(\theta) - \beta)/2]}{\sin[(kd \sin(\theta) - \beta)/2]} E_e(\theta) \tag{7.7}$$

where the last term in (7.6) has been dropped since it does not depend on the angle θ. The pattern for an array of isotropic elements ($E_e(\theta) = 1$ for all θ) is plotted in Figure 7.5 for array sizes of 3, 6, and 12 elements. There is no phase shift between elements, and the elements are spaced a half-wavelength apart. There are several factors to note [3]. First the beamwidth decreases with increasing array size. (The actual beamwidth depends upon the element spacing.) There is one main beam and $N - 2$ sidelobes. The peak sidelobe level decreases with increasing array size, reaching a limiting value of −13.2 dB.

An array with all elements equally excited is called a uniform array. As will be discussed later, a uniform array has the narrowest beamwidth and, consequently, the highest directivity, for a given number of elements. It also has the highest sidelobe level with the peak becoming −13.2 dB as the array size gets large.

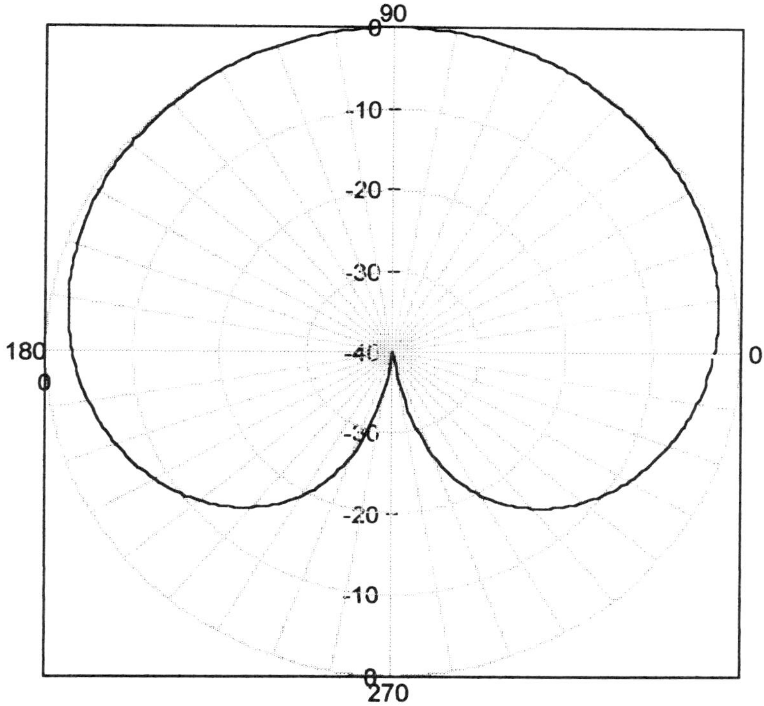

Figure 7.4 Two-element, 90-degree phase difference, equal amplitude excitation (quarter-wavelength spacing).

The position of the main beam can be moved or steered by introducing a phase shift between elements. The maximum value of (7.7) occurs when $kd\sin(\theta) - \beta = 0$. If the main beam is to be located at $\theta = \theta_0$, then the phase shift must be

$$\beta = kd\sin(\theta_0) = 2\pi\frac{d}{\lambda_0}\sin(\theta_0) \tag{7.8}$$

When using a phase shift to steer the beam, the end element is set to the reference phase, say 0 degrees; the next element is phase shifted by $-\beta$; the one after by -2β, and so on. Figure 7.6 presents patterns for a 12-element array of half-wavelength spaced isotropic elements with two different phase shifts. For purely linear phase shifts, that is equal phase shifts between adjacent elements, the pattern ''slides'' over with increasing phase shift. The larger the phase shift, the larger the beam steer. In Figure 7.6, a –20-degree phase shift (per element) steers the beam over to about 6.4 degrees, and a phase shift of –60 degrees produces a 19.5-degree beam movement.

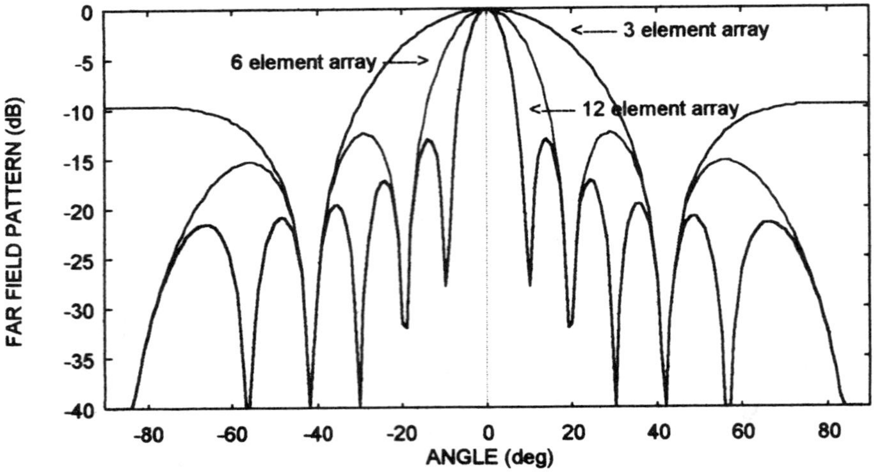

Figure 7.5 Radiation patterns for various size uniform arrays.

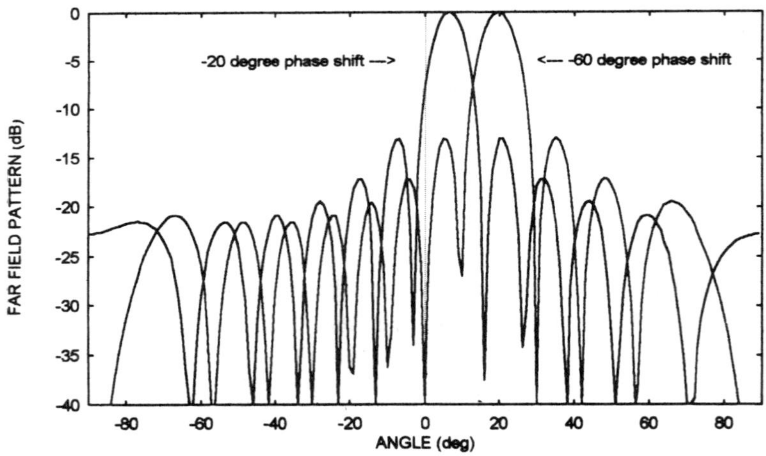

Figure 7.6 Twelve-element radiation pattern with two-element phase shift values.

The array pattern, as given by (7.7), is periodic with a period of 2π. The pattern can only be measured at angles between 0 degrees and 360 degrees. This is called the visible space. Angles greater or less than this range constitute the invisible space. A consequence of the periodicity of (7.7) is that for certain element spacings, multiple main beams appear in the visible space. Figure 7.7 shows an example of this for the 12-element

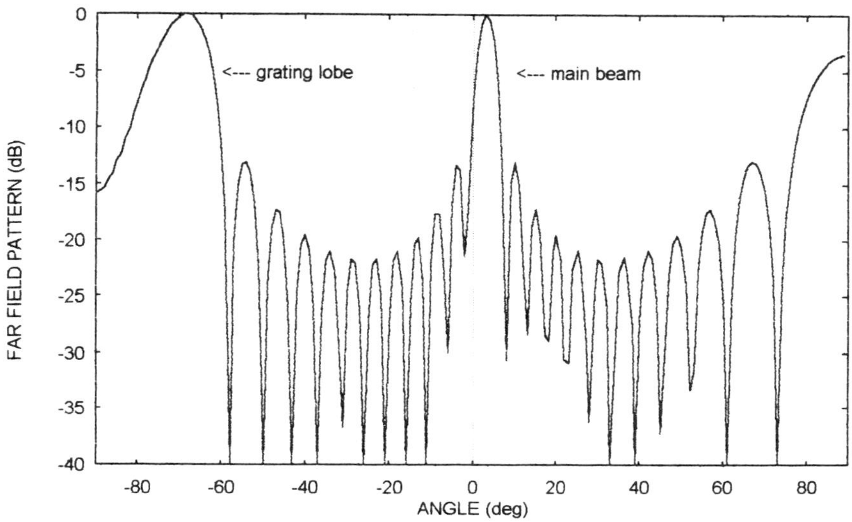

Figure 7.7 Twelve-element array pattern with a $1.016\lambda_0$ spacing and a –20-degree phase shift.

array. The element-to-element phase shift is –20 degrees, but this time the spacing is $1.016\lambda_0$. Note the second main beam occurring near –70 degrees. This second beam is called a grating lobe. Usually the presence of a grating lobe is undesirable because it becomes impossible to distinguish if a signal comes from the main beam or grating lobe.

For a given main beam angle, θ_0, a grating lobe will not appear if the element spacing, d, does not exceed

$$d = \frac{\lambda_0}{1 + \sin(\theta_0)} \tag{7.9}$$

In Figure 7.7, the spacing is larger than the maximum value, which from (7.9) is $0.935\lambda_0$. The maximum allowed spacing becomes smaller as the main beam angle increases. Grating lobes occur for any main beam angle including 0 degrees where they appear for spacings greater than a wavelength.

An array with a uniform excitation produces the narrowest possible beamwidth along with the highest sidelobe level. Sometimes it is necessary to reduce the sidelobes. High sidelobes can increase interference or result in spurious signal reception. The sidelobe level is reduced by introducing a taper in the amplitudes of the elements. When tapering the amplitude distribution, the excitation is highest at the center of the array and then decreases as one moves toward the edge. If the array has an odd number of elements, then the center element has the largest excitation. For an even number, the two elements adjacent to the center share the largest excitation.

As an example, consider an amplitude that varies linearly from a peak value of 1 at the array center to a value of C ($C < 1$) at the edge. This is shown in Figure 7.8. Although the function describing the taper is continuous, only its values at the element locations are used. In the figure, the elements are located at the dots. For the distribution in the figure

$$a_n = C + (1 - C)\left[1 - \frac{2z_n}{L}\right] \tag{7.10}$$

where a_n is the amplitude of the nth element, which is located at z_n. This type of distribution is often called a linear taper on a pedestal. The pedestal referring to the fact that the taper does not go to zero at the array ends but rather to the value C. The value C is the pedestal.

Figure 7.9 plots patterns for a 12-element array: one with a uniform distribution and one with a linear taper having a –6-dB pedestal. Since excitations refer to voltage or current, use 20 · log when converting to decibels. The two major consequences of a taper are a lowering of the sidelobe levels and the broadening of the beamwidth. These two phenomena are always linked. If a lower sidelobe level is needed, then a wider beamwidth must be accepted. Since the beamwidth widens, the directivity decreases. A significant effort has been put into finding amplitude distributions that minimize the beamwidth for a given sidelobe level or optimize the pattern in some other way. Two of the most well known are the Dolph-Chebyshev and Taylor distributions. The discussion of these is beyond the scope of this book. The reader is referred to [3] and [4], which provide excellent discussions of these amplitude distributions. For a large number of applications, the pattern requirements can be met without using sophisticated distributions.

Arrays can be two-dimensional as well. Only planar arrays will be considered here. Figure 7.10 presents the geometry for a two-dimensional, planar array. The elements are arranged in a rectangular grid defined by a pair of (x-y)-axes. The element separation and number along the x-axis may be different from that along the y-axis. The array now affects the pattern in both the θ- and ϕ-directions. When mutual coupling is negligible, pattern multiplication can be used to find the far-field pattern. Equation (7.5) becomes

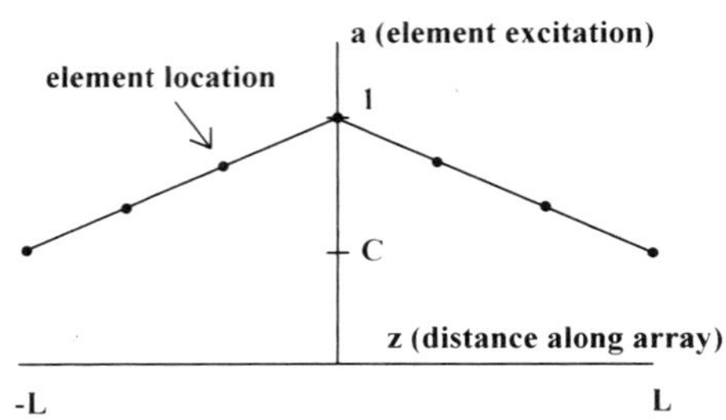

Figure 7.8 Linear tapered amplitude distribution.

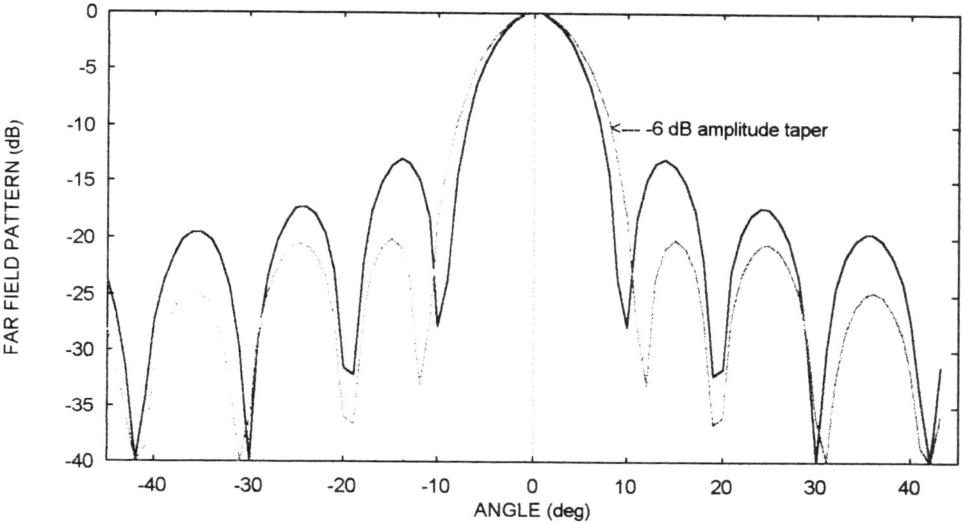

Figure 7.9 Effect of amplitude taper on a 12-element array pattern..

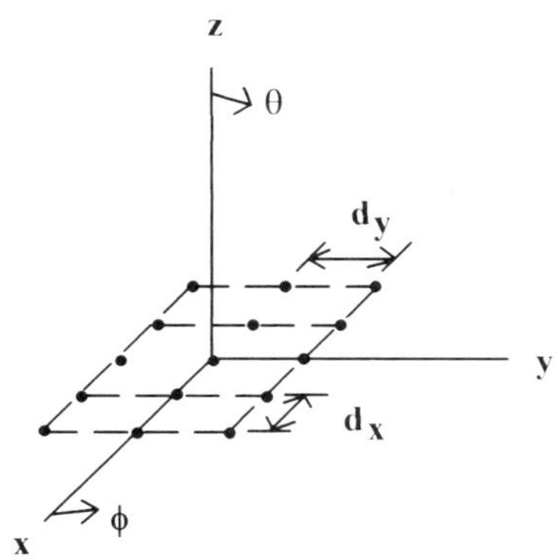

Figure 7.10 Geometry for a two-dimensional array.

$$E(\theta,\ \phi) = E_e(\theta,\ \phi)\sum_{m=1}^{M}\sum_{n=1}^{N} a_{mn} e^{j\{k\ \sin(\theta)[md_x \cos(\phi)+nd_y \sin(\phi)]-\beta_{mn}\}} \tag{7.11}$$

where the element excitation is $a_{mn}e^{-j\beta_{mn}}$. The number of elements in the x- and y-directions are M and N, respectively. In (7.11), the term containing the r dependency has been dropped.

In many situations, the distribution along the rows (say the x-direction) is identical, with each row having a different amplitude and phase. Such a distribution is called separable. The element excitation is expressible as $a_m a_n e^{-j\beta_m} e^{-j\beta_n}$. Equation (7.11) now becomes

$$E(\theta, \phi) = E_e(\theta, \phi) \sum_{m=1}^{M} a_m e^{j[mkd_x \sin(\theta)\cos(\phi) - \beta_m]} \sum_{n=1}^{N} a_n e^{j[nkd_y \sin(\theta)\sin(\phi) - \beta_n]} \tag{7.12}$$

Finally if the phase shifts are linear along x and y (that is, $\beta_m = (m - 1)\beta_x$, for example), then

$$E(\theta, \phi) = E_e(\theta, \phi) \sum_{m=1}^{M} a_m e^{j\{(m-1)[kd_x \sin(\theta)\cos(\phi) - \beta_x]\}} \sum_{n=1}^{N} a_n e^{j\{(n-1)[kd_y \sin(\theta)\sin(\phi) - \beta_y]\}} \tag{7.13}$$

Notice that (7.13) is the product of the element pattern and two array factors—one along x and one along y. The advantage of a separable distribution is that it allows one to design a planar array by considering it to be two linear arrays.

The phase shifts are used to steer the beam just as with a linear array. To position the main beam at θ_0, ϕ_0, the required phase shifts are

$$\beta_x = kd_x \sin(\theta_0)\cos(\phi_0) \tag{7.14}$$

and

$$\beta_y = kd_y \sin(\theta_0)\sin(\phi_0) \tag{7.15}$$

All of the discussion regarding grating lobes, beamwidth, and sidelobes for a linear array also applies to the planar two-dimensional array.

In any real array, the actual amplitude and phase distribution will differ from the design values. Errors will occur for a variety of reasons. The radiating element impedances will not be exactly as expected because of, for example, variations in construction, limitations of the design theory, and mutual coupling. The feed network also contains errors for the same reasons. There are two types of excitation errors, deterministic and random, both of which cause changes in the pattern. Deterministic amplitude errors tend to fill in the pattern nulls and raise the overall sidelobe level. The effect of the phase error depends upon its distribution. The error is even if it has the same sign (and approximately the same amplitude) as one moves away from the center of the array in either direction. An example is given in Figure 7.11(a). With even errors, nulls fill in and sidelobes rise. If the error is severe enough, the sidelobes adjacent to the main beam begin to merge with the main beam. This creates *shoulders* in the main beam. Odd-phase error distributions have opposite signs on each side of the array as in Figure 7.11(b). If the curve in Figure 7.11(b) was a straight line, this would be the distribution needed to steer

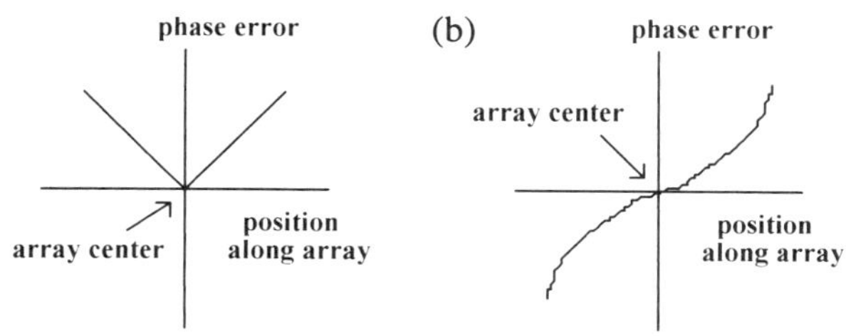

Figure 7.11 Examples of phase errors for an array excitation: (a) even-phase error and (b) odd-phase error.

the beam. When the curve is anything other than a line, the beam is also steered but, in addition, the sidelobes on the side where the beam is steered are increased while those on the opposite are decreased. Nulls fill in as well.

Random errors introduce more subtle effects. Assume that the errors are Gaussian with zero mean. Also assume that the amplitude and phase errors are independent. If the pattern with no errors is $E_0(\theta, \phi)$, then when random errors are present [5]

$$E(\theta, \phi) = \sqrt{E_0^2(\theta, \phi) + \frac{\sigma^2}{G_0}} \tag{7.16}$$

where $\sigma^2 = \sigma_a^2 + \sigma_p^2$, σ_a is the variance of the amplitude errors, σ_p is the variance of phase errors, and G_0 is the antenna gain without errors. Equation (7.16) shows that a constant amount is added to the pattern that mainly manifests itself as a small rise in sidelobes. The effect decreases as the antenna gain increases. The gain is reduced as well by

$$G = \frac{G_0}{1 + \sigma^2} \tag{7.17}$$

Note that the gain reduction does not depend on antenna size.

7.3 ARRAY CALCULATIONS AND ANALYSIS

When designing arrays it is often useful to have a quick way of estimating array performance. This helps in determining initial design parameters such as number of elements, element spacing, and excitations. Several approximations have been developed [1,2] to this end. For a linear array with a uniform excitation, the beamwidth is given by

$$\theta_{3\,\text{dB}} = \cos^{-1}\left[\sin(\theta_0) - 0.443\frac{\lambda_0}{L}\right] - \cos^{-1}\left[\sin(\theta_0) + 0.443\frac{\lambda_0}{L}\right] \tag{7.18}$$

where θ_0 is the main beam pointing angle; L is the total array length, $(N-1)d$; N denotes the number of elements; and d is the element spacing. Equation (7.18) applies to scan angles from broadside, $\theta_0 = 0$ degrees, to just short of endfire, $\theta_0 = 90$ degrees. If the array has an amplitude taper, the beamwidth broadens. For some specific distributions, (7.18) can be multiplied by a beam broadening factor [1], which accounts for the taper. These are included in a program to be described shortly. Unless an extreme taper is used, the increase in beamwidth is no more than about 20%. Equation (7.18) thus still provides a reasonable estimate when doing array sizing.

The directivity of a linear array can be estimated from

$$D = \frac{2L}{\lambda_0} = \frac{101.5}{\theta_{3\text{ dB}}} \tag{7.19}$$

which shows that it is directly related to the array length and inversely proportional to the beamwidth. When evaluating (7.19), $\theta_{3\text{ dB}}$ is in degrees. Equation (7.19) is strictly only valid for arrays with half-wavelength spacings and uniform illuminations. It is a good approximation for other spacings as well. Fortunately, directivity is a slowly varying function of the taper, so (7.19) is still useful for initial directivity estimates with other distributions.

Predicting two-dimensional array performance is somewhat more difficult. Assume an array with a separable distribution. Also assume the array is steered to the angles, θ_0, ϕ_0. Let θ_x be the beamwidth of a linear array having the same distribution as the planar array does in the x-direction (as found from (7.18)). Similarly define θ_y as the beamwidth for the distribution in the y-direction. The beamwidth in the $\phi = \phi_0$ plane is [2]

$$\theta^{-2} = \cos^2(\theta_0)\left[\frac{\cos^2(\phi_0)}{\theta_x^2} + \frac{\sin^2(\phi_0)}{\theta_y^2}\right] \tag{7.20}$$

In the plane perpendicular, the beamwidth is

$$\phi^{-2} = \frac{\sin^2(\phi_0)}{\theta_x^2} + \frac{\cos^2(\phi_0)}{\theta_y^2} \tag{7.21}$$

The directivity is found from

$$D = \pi D_x D_y \cos(\theta_0) = \frac{32400}{\theta\phi} \tag{7.22}$$

where D_x and D_y are the directivities of the equivalent linear arrays (7.19) in the x- and y-directions. θ and ϕ (both in degrees) are given by (7.20) and (7.21).

Nothing has been said about sidelobe levels. Sidelobe levels have been predicted for many distributions. Usually these are contained in tables. In lieu of tables, a program will be discussed later that calculates the pattern for a group of user-selected distributions. The program can be used to determine the distribution that gives the required sidelobe level and beamwidth.

The program ARRAYCAL provides estimates for beamwidth and directivity for both linear and planar arrays. Three different aperture distributions are built in to the program, namely uniform, cosine on a pedestal, and Chebyshev. As mentioned earlier, the uniform has the narrowest beamwidth and highest gain of any distribution. The cosine on a pedestal is expressed as

$$a_n = C + (1 - C)\cos\left[\pi \frac{z_n}{L}\right] \tag{7.23}$$

This distribution results in lower sidelobes than the uniform with a modest increase in beamwidth. The beamwidth increase is not as large as for more dramatic tapers such as the linear taper of Figure 7.8. The results from the cosine distribution are representative of those from a large group of nonoptimum but widely used distributions. The Chebyshev can be shown to give the narrowest possible beamwidth for a given sidelobe level. Although the implementation of a Chebyshev is not covered in this book, it is included in the program to give an idea of the optimum that can be obtained.

ARRAYCAL is useful for performing array sizing and the initial design. If performance specifications are known, the program will help determine the number of elements, element spacing, and aperture distribution needed to meet the requirements. Conversely the program will help find the array performance given parameters such as the number of elements and spacings. The following is an example of the program usage for a 12-element linear array at 1.5 GHz.

```
          ARRAYCAL.V10        9/11/95        13:48:10
               [Current Program Date - 12/5/95]

      Beamwidth and Directivity Analysis for Linear and Planar Arrays

ANALYZE LINEAR (l) OR PLANAR (p) ARRAYS? l

INPUT FREQUENCY (GHz)? 1.5

INPUT ELEMENT SPACING (cm)?
[wavelength =  20.000 (cm)]
?10.000

INPUT NUMBER OF ELEMENTS? 12

INPUT APERTURE DISTRIBUTION
     u - UNIFORM
     c - COSINE-ON-A-PEDESTAL
     t -  CHEBYSHEV
?c
```

For the cosine on a pedestal, the program needs to know the pedestal value, *C*. With the Chebyshev, the maximum sidelobe level must be input.

INPUT EDGE TAPER (+dB)? **6**

(This is the pedestal value. Its value in decibels (20 log) is required. Although the pedestal value is less than 1, it is input as a positive number of decibels.)

INPUT MAIN BEAM SCAN ANGLE
[broadside = 0 degrees]
?0

The beamwidth and directivity are now calculated with the following displayed on the screen.

ARRAYCAL.V12 09-11-1995 20:22:19

LINEAR ARRAY ANALYSIS

NUMBER OF ELEMENTS = 12
ELEMENT SPACING = 10.000 (cm)
FREQUENCY = 1.5000 (GHz)
COSINE-ON-A-PEDESTAL DISTRIBUTION
AMPLITUDE TAPER = 6.00 (dB)
MAIN BEAM SCAN ANGLE = 0.00 (deg)

BEAMWIDTH = 9.05 (deg)
DIRECTIVITY = 10.50 (dB)

The output displays the original input data plus the beamwidth and directivity. As usual the results can be printed out or stored in a user-named file. Should the results not be satisfactory, the program can be rerun with new inputs until the desired output is obtained. To narrow the beamwidth, either increase the number of elements or the element spacing (keeping in mind the need to avoid grating lobes, (7.9)). If this proves unsuccessful (for example, the array gets too large for the available area), one possibility is to increase the operating frequency. The beamwidth is proportional to the length of the antenna as measured in terms of wavelength. For a fixed length, the beamwidth decreases with increasing frequency. Unfortunately for many applications the frequency is fixed by other factors and not within the control of the antenna designer, so this option is typically not available.

For the planar array, the elements are assumed to lie along a grid with lines parallel to the x- and y-axes as in Figure 7.10. The distribution is assumed separable. The inputs are the same except that the number of elements, element spacing, and aperture distribution particulars (for example, type and taper) must be input for both the x and y distributions. The output for a sample case is shown below.

ARRAYCAL.V12 09-11-1995 20:26:50

PLANAR ARRAY ANALYSIS

FREQUENCY = 1.5000 (GHz)
NUMBER OF ELEMENTS ALONG x AXIS = 12
ELEMENT SPACING = 10.000 (cm)
COSINE-ON-A-PEDESTAL DISTRIBUTION
AMPLITUDE TAPER = 6.00 (dB)
NUMBER OF ELEMENTS ALONG y AXIS = 12
ELEMENT SPACING = 10.000 (cm)
COSINE-ON-A-PEDESTAL DISTRIBUTION
AMPLITUDE TAPER = 6.00 (dB)
MAIN BEAM SCAN PLANE ANGLE 0.00 (deg)
MAIN BEAM SCAN ANGLE 0.00 (deg)

SCAN PLANE BEAMWIDTH = 9.05 (deg)
PERPENDICULAR PLANE BEAMWIDTH = 9.05 (deg)
DIRECTIVITY = 25.97 (dB)

The scan plane beamwidth refers to (7.20), and the perpendicular plane beamwidth is (7.21). Although not so for this case, the parameters for the y-axis distribution can be different than those for the x-axis.

ARRAYCAL does not address sidelobe levels, nor does it provide the excitation coefficients. This is done by APERDIST. Once the basic array sizing has been performed, APERDIST can be used to adjust the distribution to obtain the needed sidelobe levels and determine the element excitations. The program only does linear arrays, but this is adequate for planar ones as well, since most planar arrays use separable distributions.

There are five aperture distributions built in to APERDIST. One is the uniform that was discussed previously. Another is the linear taper on a pedestal as seen in Figure 7.8 and given by (7.10). A third is the cosine on a pedestal (7.23). The fourth is a cosine squared on a pedestal where the excitations are

$$a_n = C + (1 - C)\cos^2\left[\pi \frac{z_n}{L}\right] \tag{7.24}$$

The final distribution is a quadratic taper on a pedestal

$$a_n = C + (1 - C)\left[1 - \frac{2}{L} z_n^2\right] \tag{7.25}$$

These distributions should be adequate for most applications. The criteria for selection is that the distribution results in the correct beamwidth and sidelobe level. Often several of these can meet a given set of specifications.

In general, it is best to choose the distribution with the minimum taper while still meeting the sidelobe requirements. By using the minimum taper, the beamwidth stays small. This keeps the directivity high. Of the five, the quadratic and cosine squared have the least taper. The one with the greatest taper is the linear. For each (except the uniform), the sidelobe level is controlled by the choice of the pedestal value, C.

After APERDIST determines the excitation coefficients, it can, at the user's option, calculate and display the resulting pattern. The pattern can be checked to see if it meets the requirements. If it does, the aperture distribution is appropriate for the antenna. If not, another pedestal value or distribution type may be tried. When determining an aperture distribution, it is always a good idea to over-design the array. The number of elements and/or element spacing should be chosen to theoretically produce a narrower beamwidth than required. Similarly the array distribution should be designed for a lower sidelobe level than needed. Real world factors such as manufacturing variations, mutual coupling, and limitations in the design equations almost always result in a wider beamwidth (and consequently lower directivity) and higher sidelobes.

To illustrate its use, the pattern of a 12-element array at 1.5 GHz with a cosine on a pedestal will be found. After invoking APERDIST, the screen looks like

APERDIST.V10 09-12-95 08:24:55
[Current Program Date - 12/4/95]

Calculates excitation coefficients for a linear array with selected aperture distributions. Calculations resulting far-field pattern.

INPUT FREQUENCY (GHz)? **1.5**

INPUT ELEMENT SPACING (cm)
[wavelength = 20.000 (cm)]
?3.937

INPUT NUMBER OF ELEMENTS? **12**
(maximum number of elements is 50)

When the number of elements is even, two options are available. One normalizes the nth/2 and (n/2 + 1st) elements to one. In this case, these would be the 6th and 7th elements. When the number is even, no element is at the geometric center of the array. Since the distribution is tapered from the center outward, the two elements adjacent to the center will have the highest values, but they will be less than one. It usually makes sense to set these element excitations to one. The other option leaves the middle two elements at the value calculated directly from the governing equation. The next input request is

NUMBER OF ELEMENTS IS EVEN
NORMALIZE n/2 AND (n/2+1) EXCITATIONS TO 1 (y or n)? **y**

If the number of elements is odd, the previous input is bypassed.

INPUT DESIRED APERTURE DISTRIBUTION
u - UNIFORM
l - LINEAR TAPER ON A PEDESTAL
q - QUADRATIC TAPER ON A PEDESTAL
c - COSINE ON A PEDESTAL
s - COSINE SQUARED ON A PEDESTAL
?c

For all but the uniform distribution, the next input is

INPUT EDGE ILLUMINATION (in positive dBs)? **6**

(As with ARRAYCAL, even though the taper is a number less than one, it is input as a positive number in decibels, again 20 log.)

APERDIST determines the element positions from the array length and number of elements. The equation for the distribution is evaluated at each element position. It then calculates the far-field pattern. To do this, the following inputs are requested.

INPUT ELEMENT TO ELEMENT PHASE SHIFT (deg)? **0**

(This input is included to allow for beam steering.)

INCLUDE ELEMENT PATTERN (y or n)? **n**

When answered no, the next input is

INPUT START AND STOP ANGLES FOR PATTERN CALCULATION
(broadside = 0 degrees)
?−45,44

(The pattern can be plotted for three different angle scales, namely, ±45 degrees, ±90 degrees, and ±180 degrees. To better see the beamwidth and sidelobe level, the smallest scale is probably the best to use.)

INPUT NUMBER OF ANGLES TO CALCULATE? **89**

APERDIST calculates the pattern and from this finds the main beam angle. This is printed out along with a request to plot the pattern as follows.

SCAN ANGLE = 0.0 deg
WANT TO PLOT PATTERN (y or n)? **y**

RECTANGULAR (r) OR POLAR (p) PLOT? **r**

The rectangular plot is probably more suitable when looking for beamwidth and sidelobe values. As shown in the following, its size can be varied whereas the polar plot covers a complete 360-degree angle range with only one amplitude range (0 to −40 dB). When answered **r** for rectangular

INPUT dB RANGE [20, 40, 60, 80] FOR PLOT? **20**

(This is the scale for the pattern values. Again when checking out the beamwidth and sidelobe level, the 20-dB range, assuming the maximum sidelobe is greater than 20 dB, is probably a good choice.)

INPUT ANGLE SCALE [45, 90, 180]? **45**

(This coincides with the angles input earlier for the pattern calculation. Also with this smallest scale, the beamwidth is easier to determine from the screen plot.)

The pattern is displayed on the screen as in Figure 7.12. The beamwidth is of the order of 11 degrees ((7.19) predicts about 9.2 degrees for a uniform illumination), and the peak sidelobe is about −18 dB.

?

The question mark is a reminder that the program is waiting for the enter key to be struck to continue. After hitting the enter key, the user is given the option to store the pattern in a file. The screen is cleared, and the pertinent input data and resulting excitation are displayed next.

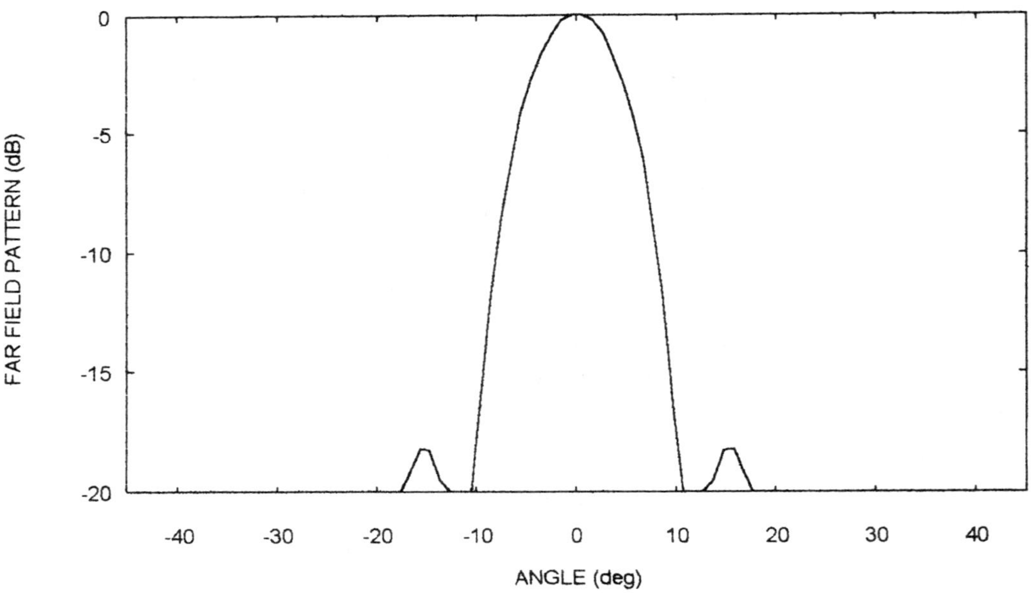

Figure 7.12 Pattern for 12-element array with a cosine on a pedestal (6-dB) distribution.

APERDIST.V10 09-11-1995 22:07:34

LINEAR ARRAY EXCITATION COEFFICIENTS

FREQUENCY = 1.500 GHz
ARRAY LENGTH = 109.982 cm
NUMBER OF ELEMENTS = 12
APERTURE DISTRIBUTION = COSINE ON A PEDESTAL

ELEMENT	POSITION (cm)	AMPLITUDE	PHASE (deg)
1	−54.991	0.501	0.00
2	−44.994	0.655	0.00
3	−34.994	0.794	0.00
4	−24.996	0.905	0.00
5	−14.999	0.976	0.00
6	−4.999	1.000	0.00
7	4.999	1.000	0.00
8	14.999	0.976	0.00
9	24.996	0.905	0.00
10	34.994	0.794	0.00
11	44.994	0.655	0.00
12	54.991	0.501	0.00

The element position, amplitude, and phase are presented. The latter is calculated from the input element-to-element phase shift. The data can be stored in a file. Two formats are possible. One stores the data just as displayed on the screen, which is useful for recording the results. The other stores only the position, amplitude, and phase data, each separated by a comma. This file may be used to replot the pattern using NARRAYD, which is discussed later. The final input asks if another case is to be done.

The element pattern can be included in the array calculations. This is done by answering yes to the appropriate input request. To include the element pattern, the pattern must be in a file. The file consists of two columns of data with the first number being the pattern angle in degrees. This is followed by a comma and then the pattern value in decibels. The additional inputs are

INPUT NUMBER OF ANGLES TO BE CALCULATED? **90**

(This tells APERDIST how many pattern values to expect. The array pattern is calculated only at the angles for which element pattern values are available.)

INPUT FILENAME CONTAINING ELEMENT PATTERN
[File must be of the form; angle (deg), field (dB)]
?c:\data\ele.pat

Some of the programs described in Chapter 3 can be used to generate element patterns for certain patch antenna structures.

The combination of programs ARRAYCAL and APERDIST plus (7.19) to (7.23) provide the necessary tools for the first part of an array design. Factors such as number of elements, element spacings, and excitation coefficients can be found to meet design specifications. The next step in the array design is to decide on an implementation.

Before discussing array feeds, one other program will be covered. NARRAYD is a general-purpose array pattern calculating program. It uses (7.5) to find the far-field pattern of an arbitrary array including nonuniform element spacings, nonlinear phase shifts, and any type of amplitude distribution. There are many options available and, since the operation of the program is fairly self-explanatory, an example will not be given. Instead a quick run-through of the inputs will be done. Many of the inputs are identical to those in other programs that have a pattern plotting capability.

When run, the program displays the usual header information. The first inputs are the frequency in gigahertz followed by the number of elements (50 maximum) in the array. The program then asks if a pattern normalization factor is to be input. Normally when performing the pattern calculation, the maximum pattern value is used to normalize the results so that when converted to decibels, the peak value is 0 dB. NARRAYD allows for the use of an arbitrary normalization factor. This may be useful when comparing, for example, different distributions, array sizes, and element spacings. As part of the output, NARRAYD prints out the normalization factor that was used.

Next the user has the option of entering the array data via the keyboard or a file. If the latter is chosen, the program requests the name of the file with the array data. This file contains a line of data for each element. The data is stored with three numbers per line, each separated with a comma. The number of lines in the file is equal to the number of elements in the array. The first number is the element position (in centimeters), the second is the element excitation amplitude, and the third is the element phase (in degrees). If the keyboard option is selected, the user is stepped through a series of inputs requesting the element positions, amplitudes, and phases. The exact inputs depend upon whether or not the element spacing is uniform, the amplitude is uniform, and the phase shift is linear (progressive).

After inputting the array data, NARRAYD clears the screen and displays a table of element positions, amplitudes, and phases. This gives the user a chance to review the data for correctness. The program asks if any changes are needed. Any of the values except the frequency and number of elements can be changed. The next input asks if an element pattern is to be included. If answered yes, the program requests the number of angles to be calculated and the name of a file containing the element pattern. The format of the file should be two values per line separated by a comma. The first is the angle in degrees, and the second is the pattern magnitude in decibels. The total number of lines in the file equals the number of angles for which there are element pattern values. The array pattern is only calculated at the angles contained in the element pattern file. If no element pattern is used, the start and stop angles along with the number of angles to use are requested.

NARRAYD then calculates the array pattern. When it is finished, it prints out the main beam scan angle and the normalization factor. The program then asks if the user wants the pattern plotted on the screen. The usual rectangular and polar plots are available. The rectangular plots can have amplitude ranges of 20 dB, 40 dB, 60 dB, and 80 dB. The angle scale can be ±45 degrees, ±90 degrees, and ±180 degrees. After displaying the pattern, the program waits for the enter key to be struck before continuing. Finally, the user has the option to store the pattern in a file.

7.4 ARRAY ARCHITECTURES

Once the various array parameters have been established, the next step is to design the feed network. It is the function of the feed to provide the correct element excitation and an impedance match at its input. For an array with N elements, the feed is usually an $(N + 1)$-port network. N of the ports are connected to the radiating elements, and one serves as the input/output depending upon whether the system is transmitting or receiving. There is a class of feeds that have multiple inputs. These produce simultaneous multiple beams with each beam pointing in a different direction. This type of feed is beyond the scope of this book.

Not surprisingly there are many types of feed structures. One class, called space feeds, uses free space to guide the signal from the input to the principal radiator. An example is the parabolic dish. A small feed horn is placed at the focus of the parabola. The signal radiated by the horn spreads out and is intercepted by the dish. The dish acts like an array in that it concentrates the feed signal into a very narrow pattern. Another class of feeds uses a large transmission line structure, like a pair of parallel metal plates. The feed is located at one end of the plates. The plates both guide the signal and allow it to spread out so that an array located at the opposite end can be excited. A third class of feeds, called constrained feeds, directly connects to each element with transmission lines or waveguides. Although arrays of microstrip antennas have been built with all of the above feed types, the last one is by far the most commonly used. These feeds will be discussed in this section.

With one type of constrained feed, the elements are connected in parallel. This is called a corporate or parallel feed. An example of a corporate fed linear array is given in Figure 7.13. Pairs of patches are connected with reactive power splitters. Connected pairs are, in turn, connected with another power splitter. With the four patches in the figure, only two layers of splitters are required. For larger arrays, more levels are needed. The amplitude excitations of the patches are controlled by the power splitters. In the figure, the inner patches have a higher amplitude because of the power splitter design. If the beam is to be steered, the required phase shifts are realized by making the horizontal line lengths unequal. The line lengths for the part of the feed connecting pairs of patches

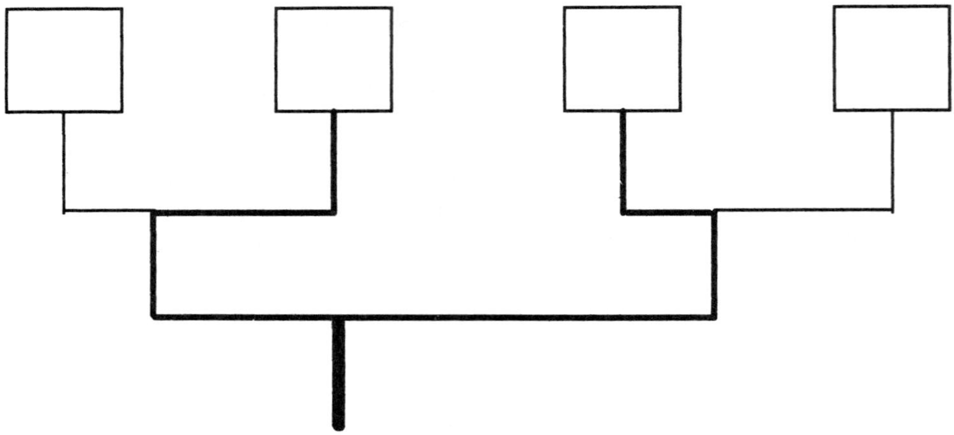

Figure 7.13 Corporate-fed linear array of microstrip patches.

are set up to produce the desired element to element phase shift between adjacent patches. The lower part of the feed produces a phase shift of twice the interelement value so that the third element has the proper phase with respect to the leftmost one.

The feed in Figure 7.13 uses reactive power splitters. Because the splitter outputs are not isolated, some care must be exercised when they are used. The patch design must be as accurate as possible to minimize impedance mismatch reflections that will disrupt the excitations. Symmetry should be exploited to counter any reflections that may occur. The use of reactive splitters works best for uniform, in-phase distributions. Here the paths to all patches are identical. Matched splitters such as Wilkinsons or branch line hybrids may also be used. Unless the pattern requirements are quite stringent (for example, very low side lobes), matched splitters are not used because they take up considerable space and require a matched load or resistor that increases cost.

Two-dimensional arrays are also constructed with corporate feeds. Figure 7.14 illustrates one example. In the figure, the patches are fed in phase with equal amplitudes, but the same approach as shown in Figure 7.13 can be used when an amplitude taper and/or phase shift are desired. Note that the array is composed of four subarrays with each having four elements. The subarray is considered the basic building block of the antenna. Large corporate arrays are often constructed by arraying subarrays [6]. The design becomes repetitive, making it easy to implement. Most corporate fed arrays consist of 2^N elements because of the two-way power splitters. The array does not have to be square. The number of elements along one side may be different from that on the other.

Corporate-fed arrays are mainly used for antennas with broadside beams where the inherent symmetry can be exploited. They are fairly straightforward to design, especially for uniform distributions. By using the modular approach described in the previous

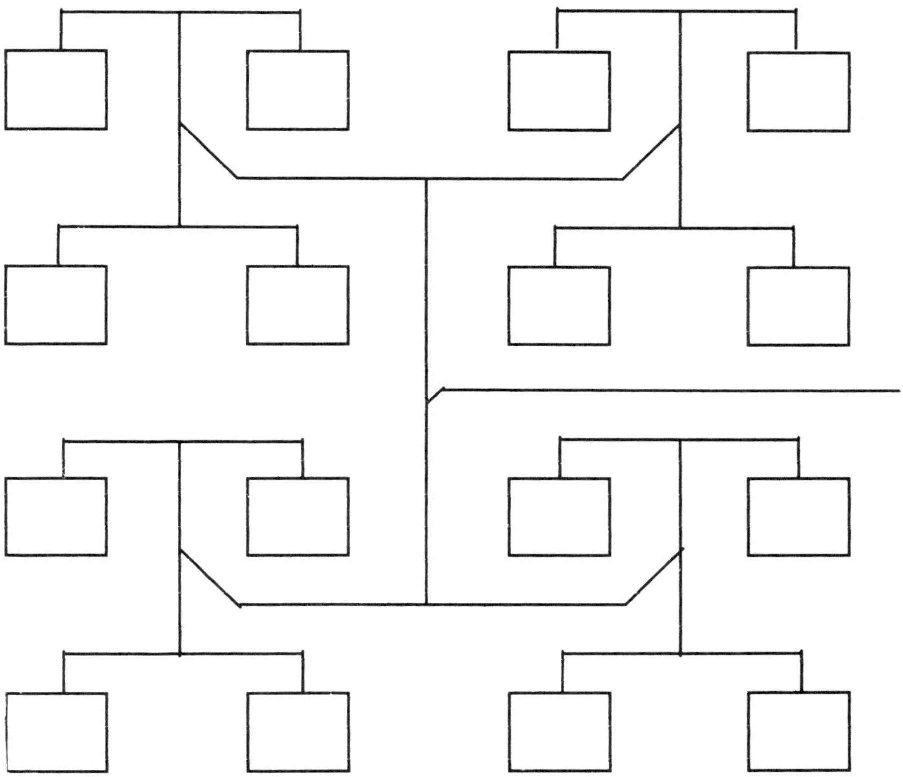

Figure 7.14 Two-dimensional corporate-fed microstrip patch array.

paragraph, different sized arrays can be assembled quite quickly. One of the drawbacks to corporate feeds is that they take up considerable room. As the array size increases, so does the line length to any element. The losses become larger, thereby reducing antenna efficiency. In fact, for these arrays a saturation effect occurs as the array size gets bigger. For a while, the antenna gain increases as it should due to the larger size, but eventually the line losses increase at a rate equal to or sometimes greater than the gain. At this point, the gain either remains essentially flat or may actually decrease with more elements. Also feed line radiation can become a problem. Not only is there some radiation from the straight sections, but more importantly, every bend and T-junction is a source of spurious radiation. The main effects of feed line radiation are increased side lobes and a small reduction gain.

Microstrip arrays may also be constructed by connecting the elements in series. Two types of series-fed linear arrays are given in Figure 7.15. The patches are interconnected with microstrip lines. The patches may be mounted in series with the feed lines as in Figure 7.15(a) or in shunt as in 7.15(b). For the series-mounted patches, the amplitude of the element excitation is controlled by varying the patch width. With the shunt-mounted elements, the width of the microstrip line from the main feed line to the patch controls the amplitude. The element-to-element phase shift is controlled by the length of the feed line between patches. After the last element is a final section of feed line and a terminating load.

There are two forms for the series feed array depending upon the main beam angle. For a broadside beam, the elements have to be in phase. This means that the phase shift between patches must be 360 degrees (or some multiple thereof). The interconnecting lines along the main feed (the horizontal lines in the figure) are thus a half-wavelength-long for Figure 7.15(a) and a wavelength-long in Figure 7.15(b). The wavelength in the microstrip is less than the free-space value. The elements are spaced less than a free-space wavelength, so no grating lobe problems arise. A series-fed broadside array is called a resonant array because of the wavelength spacing. If the beam is to be steered, the interconnect length is chosen to provide the necessary phase shift. The elements are no longer a wavelength apart. Antennas with steered beams are called nonresonant or traveling wave arrays.

Each element introduces a mismatch to the line. The amount of power an element extracts from the line is related to its impedance relative to that of the line. If one attempts to radiate all of the input power, the elements must, of necessity, become very low in impedance. The reflections induced by the low impedances become quite large and cause serious errors in the element excitations. This produces significant pattern distortions. As a result, series feeding is usually only employed for arrays with reasonable (for example, 10 or more) numbers of elements. The larger the array, the less power each element must extract from the feed line. This keeps the element mismatches low.

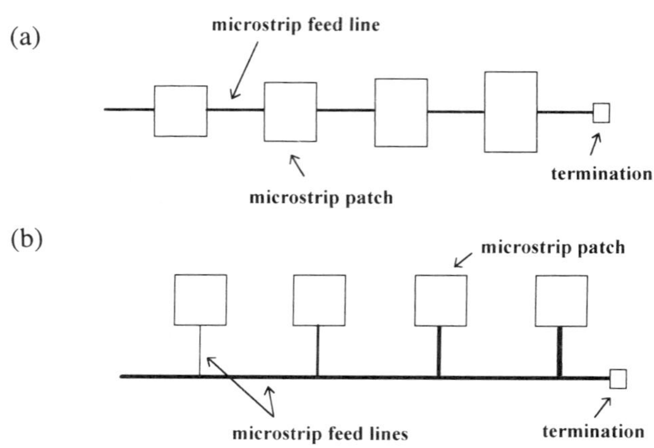

Figure 7.15 Series-fed linear arrays: (a) series-connected patches and (b) shunt-connected patches.

With traveling wave arrays, it is impossible to radiate all of the power input to the antenna. The termination is a matched load to absorb any power not radiated. If the array is not properly terminated, any residual power is reflected and travels back down the feed line in the opposite direction. It radiates with a main beam pointing at the negative of the designed main beam angle. Since the power in the reflected wave is less than the input signal, the second main beam appears as a very high side lobe.

The matched load is usually realized using a chip resistor having the same resistance as the characteristic impedance of the feed line connecting the elements. At microwave frequencies, it is important to use a good quality resistor with as small a size as possible. This minimizes the parasitics associated with the termination. One side of the resistor is connected to the feed line. Usually this is done by soldering one of the resistor pads to the feed line. The other side is connected to ground. Figure 7.16 shows two possible ways of grounding the resistor. In Figure 7.16(a), an open circuited stub is used to place an RF ground at the resistor. The stub length should be about a quarter wavelength (on the

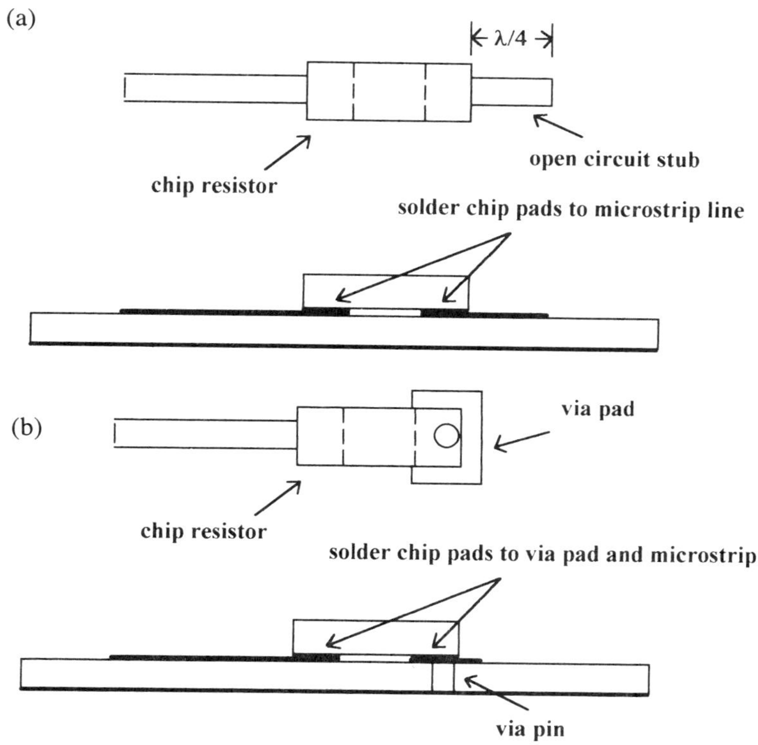

Figure 7.16 Possible layouts for terminations in a series fed array: (a) a termination using an open circuit stub and (b) a termination using a via to ground.

microstrip line) minus the open end effect extension. The other way is to use a via pin to ground as in Figure 7.16(b). In both cases, it is advisable to keep any changes in dimensions to a minimum to reduce stray capacitance and inductance. For example, the resistor and line widths should be comparable if possible. Also the via pad size should be kept small and have a width comparable to that of the resistor.

For a resonant array, the elements are a wavelength apart, so any reflected power still produces a broadside beam. To maximize the radiation, the reflected power radiation must be in phase with the incident power. This is done by either (i) eliminating the feed line and termination after the last patch (this places an open circuit at the last element), (ii) extending the feed line a half-wavelength beyond the last patch and leaving it open-circuited, or (iii) making the feed line a quarter-wavelength long and terminating the line with a short circuit.

With either type of array, the phase shift caused by the feed lines varies with frequency. For the traveling wave array, this means that the main beam angle changes with frequency (increasing with frequency). Since the phase shift increases with frequency, the beam squint angle is less than the design value below the design frequency. Above the design frequency, the beam squint is greater. Because the elements are spaced at some value that is not an integral of a wavelength, the mismatches from each travel a different distance to reach the feed point. As a result, they add up in a somewhat random fashion. This produces a relatively low SWR over a fairly wide bandwidth. The bandwidth is limited by how much beam movement can be tolerated. In a resonant array, all the mismatches travel an integral number of wavelengths and therefore add up at the feed point. Consequently, the impedance bandwidth is very narrow. The larger the array, the narrower the bandwidth. Resonant arrays can only be used over narrow frequency ranges. The changing element phase shifts appear mainly as a broadening of the broadside beam and increase in sidelobe level at frequencies on either side of the design value.

The feed point can be located at the center of the array as in Figure 7.17. With arrays having a beam squint, the feed is offset such that the path length to one side is 180 degrees longer than the other. This is done to ensure that the main beams from each half of the array are in-phase. Center feeding results in a more efficient array. Power is injected at the array center where the element excitation is highest. The power reaching the termination may be higher, however, because of fewer elements between the feed and termination. Center feeding is particularly attractive for broadside beams because it possesses maximum symmetry.

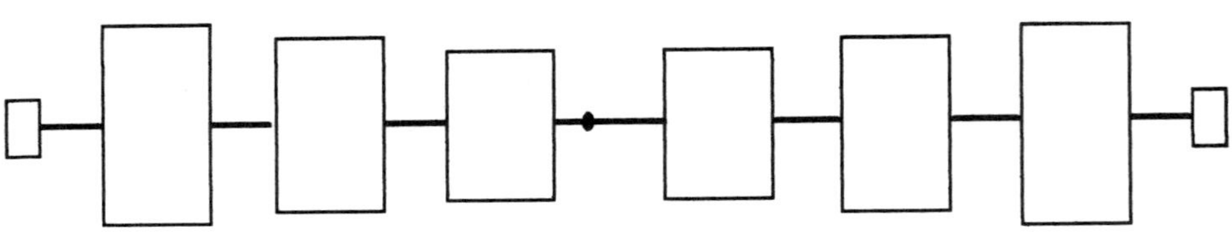

Figure 7.17 Series array with center feeding.

Two-dimensional series-fed arrays are also possible. An end-fed array is shown in Figure 7.18. The vertical series feed can be either end or center fed. In the latter, the antenna is end fed in one plane and center fed in the other. Another variant is given in Figure 7.19. Here the array is completely center fed if the feed is in the middle of the vertical microstrip line. It is a combination center-end fed when the feed is at one end of that line. Again a 180-degree offset is required for the rows on either side of the vertical center feed.

Series feeds are used when a steered main beam is required and the beam movement with frequency can be tolerated. In this situation, they are easier to design and implement compared to a corporate array. Series-fed arrays have significantly less feed line length and therefore take up less space. The reduced feed length also minimizes feed line loss and radiation. For large arrays, series feeding is typically more efficient. They are more difficult to design because the radiating elements are disbursed within the feed line. The patterns are more frequency sensitive with the sidelobes and other features changing with frequency. Finally, large resonant arrays have very narrow impedance bandwidths.

Two-dimensional arrays can have combinations of corporate and series feeds. These are sometimes referred to as hybrid feeds. Figure 7.20 presents an example. Which type of feed to use and in which plane to use them is a function of the requirements on the

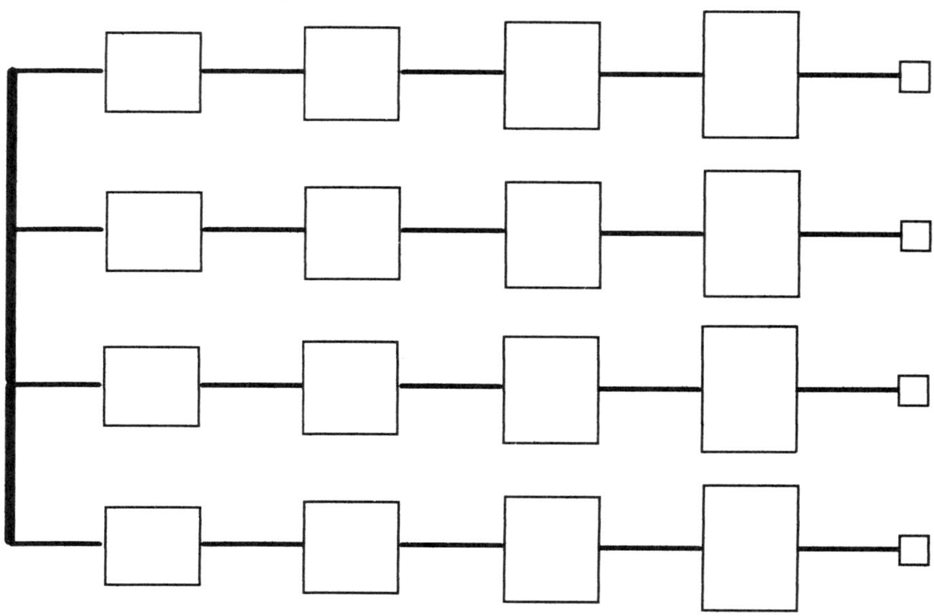

Figure 7.18 End-fed two-dimensional array.

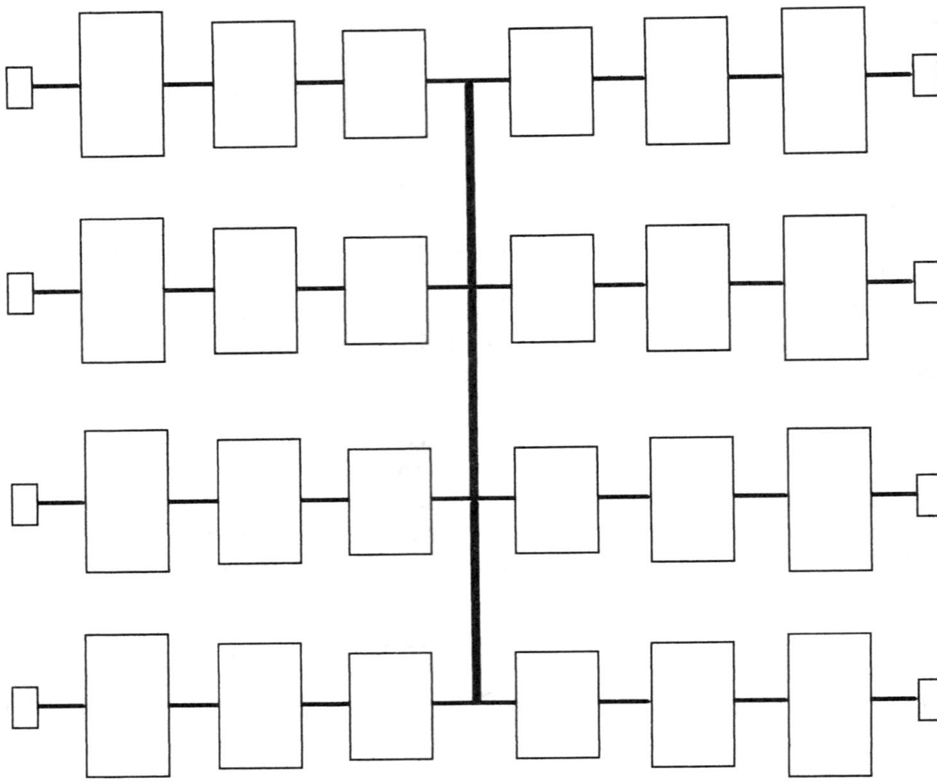

Figure 7.19 Center-fed series array.

antenna. Reference [7] has a very good discussion of both linear and two-dimensional feeds, including the advantages and disadvantages of the various types.

7.5 CORPORATE ARRAY DESIGN

In this section, a design procedure for corporate-fed arrays will be presented. As mentioned in the previous section, two-dimensional arrays are often built up from four-element subarrays. The discussion here will therefore concentrate on the subarray design. Essentially the same approach is suitable for linear arrays as well. Assume that a two-dimensional array must be designed at 1.5 GHz with the beam scanned 7 degrees in one plane and no scan (broadside beam) in the other. The beamwidth and sidelobes in the plane of scan are 16 degrees and −20 dB, respectively. The corresponding values in the other plane are 28 degrees and −10 dB. The required directivity is 18.5 dB.

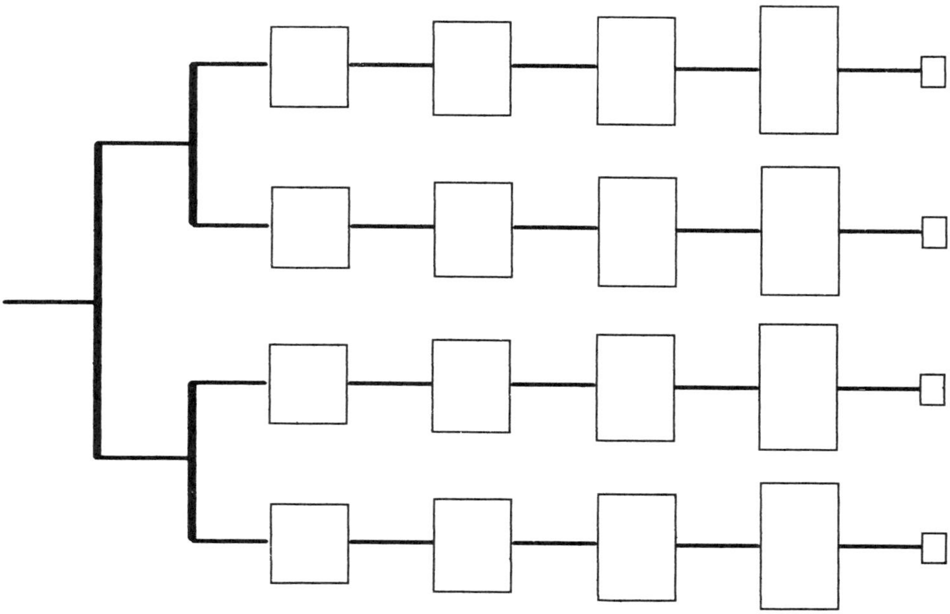

Figure 7.20 Hybrid-fed array with a combination corporate/series feed.

The first step is to estimate the number of elements needed in each plane. To do this, an element spacing must be established that, in turn, depends somewhat on the patch size. As a start, assume a square patch. While the final patch may not be square, this is still a reasonable starting point. Very narrow patches are not used because of their high input impedance. Wide patches are also not used because with typical element spacings, the patches may be too close together and result in significant mutual coupling.

The choice of substrate dielectric constant and thickness will impact the design in several ways. First the possibility of surface-wave excitation must always be considered. Higher dielectric constants result in smaller patch sizes, which can allow for closer spacings. Element spacings are expressed in terms of free-space wavelengths, so usually there is no problem fitting the elements in for normal separations. Patches on higher dielectric constant materials will have higher impedances, smaller bandwidths, and lower efficiencies. Since the element spacing is a function of the free-space wavelength, the feed line lengths will be longer, increasing losses and the chances for spurious radiation. Generally speaking, it is advisable to use the lowest possible dielectric constant material.

For this case, the substrate is chosen to be 0.16-cm thick with a relative dielectric constant of 2.2, loss tangent of 0.001, and a conductor thickness of 0.00356 cm (1 oz. copper). PATCH9 is run to get an idea as to the patch size. For the feed options, choose none for the feed type and zero inset. The resulting patch is 6.622-cm square The input impedance is found to be 279Ω.

The allowable element spacings are determined by (7.9). In the plane of scan, it is $0.891\lambda_0$. In the other plane, it is λ_0. Neither of these are very restrictive. Typical element spacings are between $0.5\lambda_0$ to $0.75\lambda_0$. With a low dielectric constant substrate, the former is probably too small ($0.169\lambda_0$ between patch edges) and may produce excessive mutual coupling [8]. The wider spacings are usually not used as well because feed line lengths become long. As a compromise for this design, a spacing of $0.65\lambda_0$ (13.000 cm) will be selected. With this value, the element-to-element phase shift in the plane of scan is –28.52 degrees (from (7.8)). The spacing between adjacent patch edges becomes 6.378 cm ($0.319\lambda_0$), which, according to [8], should keep mutual coupling low.

ARRAYCAL is run to find the number of elements in each plane by calculating the beamwidth as a function of the number of elements. Each plane is treated as a linear array, and the calculations performed assuming a broadside beam and a uniform distribution. With a 13.000-cm spacing, the beamwidths for 4, 6, and 8 elements are 26.26 degrees, 15.67 degrees, and 11.17 degrees, respectively. Keeping in mind the need to somewhat over-design the array, it appears that 8 elements is a good starting point for the scan plane and 4 elements for the other plane. APERDIST is now used to find the amplitude distribution. Several distributions will produce –20 dB or greater sidelobes. The cosine on a pedestal with a 10-dB taper is chosen for implementation. It gives a –22-dB sidelobe, which provides some design margin. The output from APERDIST for this distribution is

APERDIST.V10 09-17-1995 16:55:08

LINEAR ARRAY EXCITATION COEFFICIENTS

FREQUENCY = 1.500 GHz

ARRAY LENGTH = 90.998 cm
NUMBER OF ELEMENTS = 8
APERTURE DISTRIBUTION = COSINE ON A PEDESTAL

ELEMENT	POSITION (cm)	AMPLITUDE	PHASE (deg)
1	–45.499	0.316	0.00
2	–32.499	0.658	–28.52
3	–19.500	0.908	–57.04
4	–6.500	1.000	–85.56
5	6.500	1.000	–114.08
6	19.500	0.908	–142.60
7	32.499	0.658	–171.12
8	45.499	0.316	–199.64

The sidelobe requirement for the other plane is easily satisfied with a uniform distribution.

A 4-element array with a uniform distribution produces a sidelobe level of somewhat less than −11 dB, again giving some design margin. The excitations are

APERDIST.V10   09-17-1995   16:56:45

LINEAR ARRAY EXCITATION COEFFICIENTS

FREQUENCY = 1.500 GHz
ARRAY LENGTH = 38.999 cm
NUMBER OF ELEMENTS = 4
APERTURE DISTRIBUTION = UNIFORM

ELEMENT	POSITION (cm)	AMPLITUDE	PHASE (deg)
1	−19.500	1.000	0.00
2	−6.500	1.000	0.00
3	6.500	1.000	0.00
4	19.500	1.000	0.00

To verify the results, ARRAYCAL is run once more, this time analyzing the planar array. Since a cosine on a pedestal is available within ARRAYCAL, the results will correspond with the selected distributions. If a different distribution, such as the linear on a pedestal, is used, the cosine on a pedestal in ARRAYCAL can still be used to provide a reasonable estimate of performance. As an alternative, calculate the patterns with the chosen distributions using NARRAYD and find the beamwidth and sidelobe level from the results. If necessary, (7.20) to (7.22) can be evaluated directly. From ARRAYCAL

ARRAYCAL.V14   09-17-1995   16:58:15

PLANAR ARRAY ANALYSIS

FREQUENCY = 1.5000 (GHz)
NUMBER OF ELEMENTS ALONG x AXIS = 8
ELEMENT SPACING = 13.000 (cm)
COSINE-ON-A-PEDESTAL DISTRIBUTION
AMPLITUDE TAPER = 10.00 (dB)
NUMBER OF ELEMENTS ALONG y AXIS = 4
ELEMENT SPACING = 13.000 (cm)
UNIFORM DISTRIBUTION
MAIN BEAM SCAN PLANE ANGLE 0.00 (deg)
MAIN BEAM SCAN ANGLE 7.00 (deg)

SCAN PLANE BEAMWIDTH = 14.72 (deg)
PERPENDICULAR PLANE BEAMWIDTH = 26.47 (deg)
DIRECTIVITY = 19.20 (dB)

The array should meet or exceed the design goals. There is some design margin to help deal with design inaccuracies and fabrication tolerances. The pattern of the array in the scan plane is shown in Figure 7.21. The pattern in the other plane can be seen in Figure 7.22. Everything appears to be satisfactory, so the design may continue.

A sketch of the array with the excitations along the rows and columns is given in Figure 7.23. Since the excitation is symmetrical about the array center, only half the elements are marked. The first number in the parenthesis is the amplitude. The second is the phase. The splitters between patch pairs provide a –28.52-degree phase shift plus an unequal power division. Also shown are the phase shifts to be incorporated into the splitters connecting the subarrays in the horizontal direction. These circuits must also provide the correct power split. The power dividers connecting the subarrays in the vertical direction are all equal amplitude, in-phase splitters because of the uniform distribution.

We will concentrate on the design of the feed for the subarray in the upper left-hand corner. The power split ratio between patches is 0.658/0.316 = 2.0823:1. Applying (5.26) to (5.30) yields line impedances of $Z_1 = 3.0823Z_0$ for the arm with the lower power and $Z_2 = 1.480Z_0$ for the other arm. Z_0 is the desired input impedance of the power divider. The feed line from the input runs to the next level of the feed where it is combined again. Ideally, Z_0 should be high, like 100Ω to 200Ω, to ease in combining at the next and further down levels. The maximum microstrip line impedance is limited by the narrowest

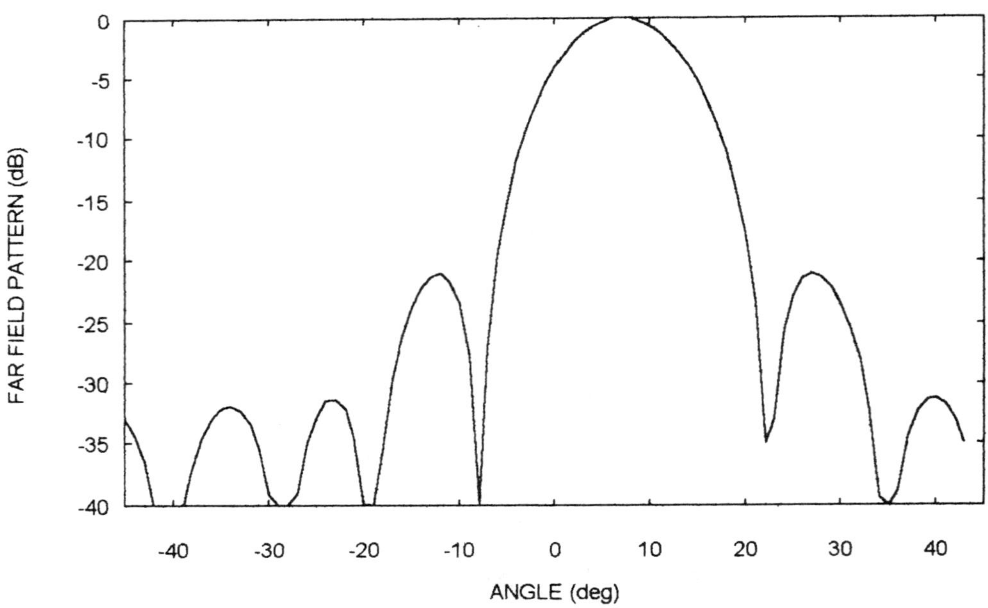

Figure 7.21 Array pattern in the plane of scan.

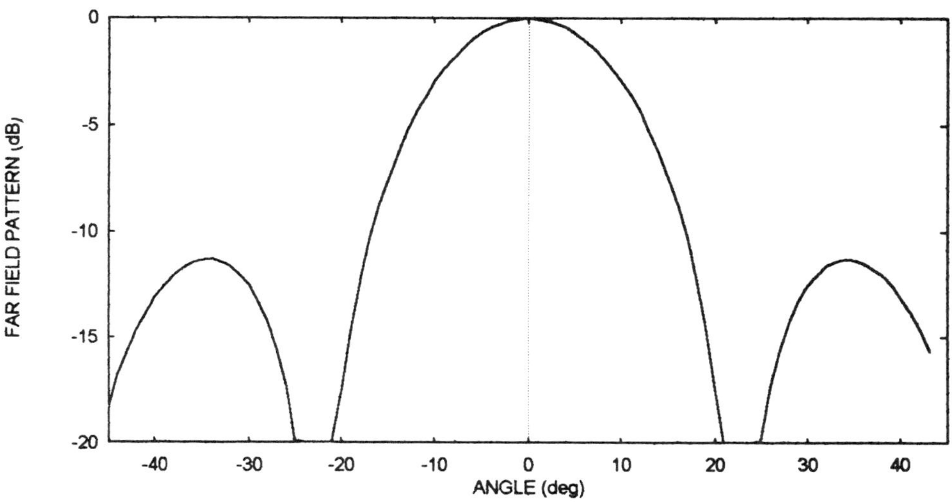

Figure 7.22 Array pattern in plane orthogonal to scan plane.

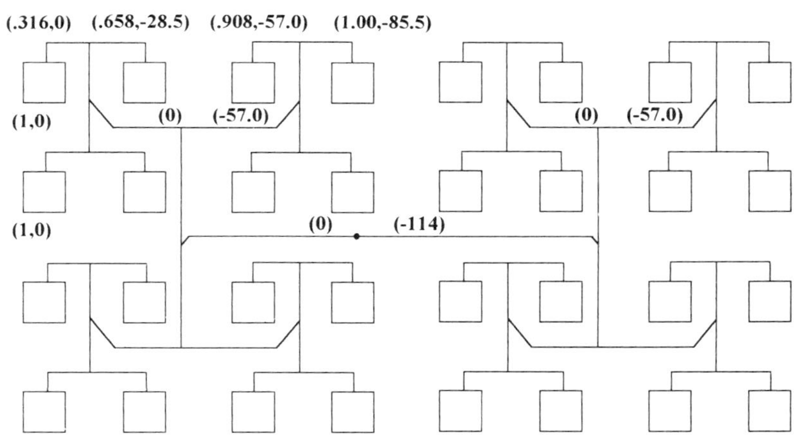

Figure 7.23 Element excitations for an 8 × 4 corporate-fed array.

line that can be accurately etched. Typically the highest impedance is around 200Ω, which for the substrate under consideration has about a 0.0127-cm line width. If Z_0 is chosen to be 100Ω or greater, Z_1 will exceed the maximum impedance value. There is thus a trade-off between a high value for Z_0 and the need to avoid extremely narrow lines. To keep Z_1 manageable, Z_0 is taken to be 50Ω. This makes Z_1 = 154.1Ω and Z_2 = 74.0Ω.

The patches must be matched to the output line impedances of the splitter. This requires quarterwave transformers at the patches with impedances of 207Ω and 143.6Ω (patch input resistance = 279Ω). The former requires a line width of 0.00889 cm, which exceeds the assumed narrowest acceptable line width of 0.0127 cm. A 0.0127-cm-wide line has a characteristic impedance of 196Ω. This would be the highest transformer impedance available. Working backward, the patch impedance should be 249Ω in order to be transformed by the 0.0127-cm line to 154Ω. This could be done by increasing the patch width or insetting the patch feed. In this case, an inset will be used if it is small. Using PATCH9 an inset of only 0.630 cm is needed. For this illustration, no correction for the gap is assumed (see Section 3.4). To convert to 74Ω for the other side of the splitter, a 135.8-Ω transformer is needed.

The feed to this point is shown in Figure 7.24. A bend will have to be put into the transformers to allow connection of the patch pair. To avoid coupling, place the bend such that the separation between the patch and the horizontal part of the transformer is around three to five substrate thicknesses. For this design, a separation of 0.762 cm is selected. The separation to choose must also consider the feed line that couples the subarray to its next neighbor. As seen in Figure 7.23, the line runs between the upper and lower patches of the subarray. This line is brought out from a point located somewhere on the vertical feed connecting the upper and lower patch pairs. To minimize interference with the top right patch, this tap point should be located below or very near the lower edge of the top patch. When routed out of the subarray, the feed line must also be kept a sufficient distance from both the upper patch and the lower patch feed line.

The vertical feed line connecting the upper and lower patch pairs is displaced slightly toward the left. This is to allow the needed phase shift between right and left patches. The 74-Ω line will be longer by −28.52 degrees than the 154.1-Ω feed. The actual lengths must also include the reference plane shifts caused by the T-junction.

At the input to the splitter, the impedance (ignoring junction effects) is 50Ω. To make the next splitter design simpler, a transformer of impedance 70.7Ω will be used to up convert the impedance level to 100Ω. This is a common practice in corporate feed design. It is usually desirable to keep the impedances high to avoid the need for very low

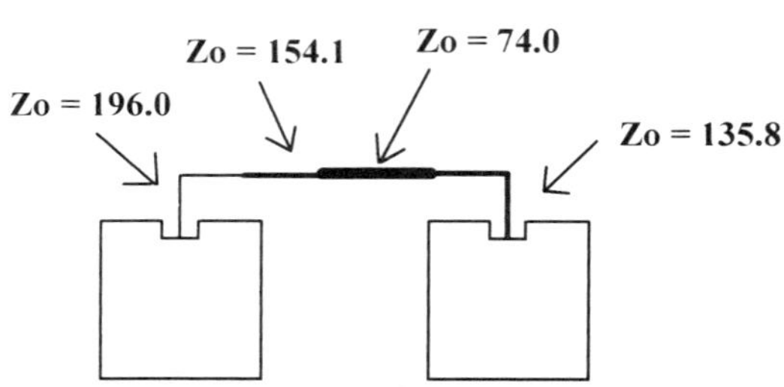

Figure 7.24 Initial design of a splitter and feed connecting two patches.

(and consequently wide) feed lines. After the transformer, a 100-Ω line will be run down to the next level splitter. With the addition of these lines, the feed now looks as in Figure 7.25. When the 100-Ω lines from each patch pair meet at the next splitter, another transformer is used. Its value depends upon the power split of the divider connecting the adjacent subarrays.

Now that the various line impedances have been determined, the next step is to find the line widths and lengths. MICRO can be used for this with the results given in Table 7.1. Wherever bends are used, optimum miters should be utilized to minimize the introduced discontinuities. The optimum chamfer is found by MMITER. The 196-Ω line

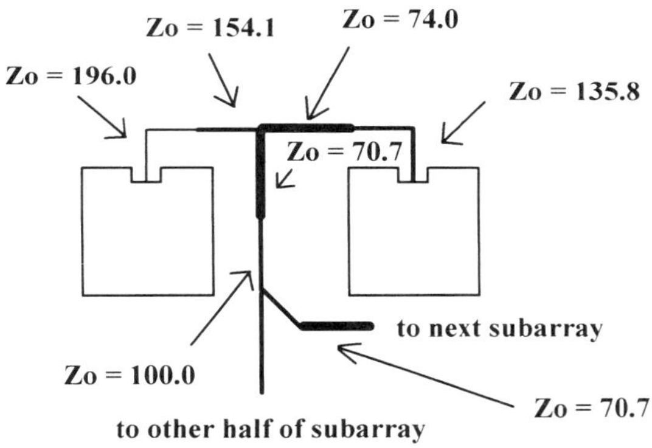

Figure 7.25 Design of subarray to second level splitter.

Table 7.1
Data From Various Microstrip Lines in the Feed Network

Impedance (Ohms)	*Effective Dielectric Constant*	*Line Width (cm)*	*Quarter Wavelength (cm)*	*Function*
196.0	1.650	0.0127	3.868	Converts patch impedance to 154.1Ω
135.8	1.706	0.0597	3.828	Converts patch impedance to 74.0Ω
154.1	1.687	0.0381	—	Output arm of splitter
74.0	1.806	0.2578	—	Output arm of splitter
70.7	1.815	0.2807	3.711	Converts splitter input impedance to 100Ω
100.0	1.753	0.1379	—	Connects to other pair of patches

width is outside the validity range of the model, but the results will still be applied. The output for both transformers are as follows.

MMITER.V10 09-20-1995 20:20:43

SUBSTRATE RELATIVE er = 2.20
SUBSTRATE HEIGHT = 0.160 (cm)
LINE WIDTH = 0.0127 (cm)
FREQUENCY = 5.000000 (GHz)

CHAMFER = 0.0140 (cm)
[Chamfer measured from outside corner of bend along outside edge of line]

REDUCTION IN ELECTRICAL LENGTH = 0.0732 (cm)
[Compared to physical length of 0.0127 (cm)]

and

MMITER.V10 09-20-1995 20:20:43

SUBSTRATE RELATIVE er = 2.20
SUBSTRATE HEIGHT = 0.160 (cm)
LINE WIDTH = 0.0597 (cm)
FREQUENCY = 5.000000 (GHz)

CHAMFER =0.0545 (cm)
[Chamfer measured from outside corner of bend along outside edge of line]

REDUCTION IN ELECTRICAL LENGTH = 0.1046 (cm)
[Compared to physical length of 0.0597 (cm)]

One side effect of using an optimum miter is that the electrical length through the bend changes. In most cases, the electrical length decreases. MMITER calculates this change. The physical length of the transformers must be increased by the difference between the electrical length reduction and the actual length. For the 196-Ω line it is lengthened by 0.0604 cm, while for the 135.8-Ω line the additional length is 0.0450 cm. The chamfer is measured as shown in Figure 7.26. The miter distance is the same on both sides of the bend. For both transformers, 0.300 in of total length is used to extend the lines out from the patches.

The 74-Ω line must include a section to produce the –28.52-degree phase shift. The wavelength on the line is $\lambda = \lambda_0/\sqrt{\epsilon_{re}} = 14.882$ cm. To obtain the phase shift, a section, $l_p = (28.52/360)\lambda = 1.179$ cm is required. Other factors to consider are the reference plane extensions and shunt susceptance at the T-junction. Line lengths are measured from the physical center of the junction, so they must correct for the reference plane shifts. These are found with MICTEE. The output from a program run gives the junction parameters. For the junction in the splitter connecting the two patches we have the following output.

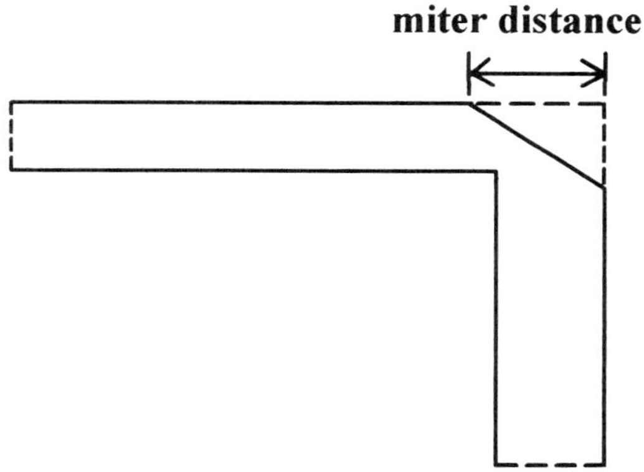

Figure 7.26 Mitered microstrip bend.

MICTEE.V11 09-20-1995 21:24:37

Microstrip T-Junction Equivalent Circuit Elements

LEFT SIDE MAIN LINE WIDTH = 0.0381 (cm)
RIGHT SIDE MAIN LINE WIDTH = 0.2578 (cm)
BRANCH LINE WIDTH = 0.2807 (cm)
SUBSTRATE HEIGHT = 0.160 (cm)
SUBSTRATE er = 2.20
CONDUCTOR THICKNESS = 0.00356 (cm)

FREQ (GHz)	d1a (cm)	d1b (cm)	nia	nib (cm)	d2 (Mhos)	Bt
1.500	0.07546	0.03599	1.00	1.00	0.1856	0.000025

With this junction, the susceptance is negligible, and the transformers have a unity turns ratio. These circuit elements may be ignored in this design.

The layout of the feed with the dimensions known so far can be seen in Figure 7.27. All dimensions are measured from the center lines of the junctions and feed lines. The corrections for the electrical lengths of the bends have been incorporated into the dimensions. Also included is the reference plane extension for the splitter input transformer.

What must be determined next are the lengths of the splitter lines, which are not dimensioned in the figure. This is done by setting up two equations—one related to the physical length and the other to the electrical length. Let l_l be the length of the 154.1-Ω section on the left. Let l_r be the length of the 74.0-Ω line on the right. The total distance between vertical sections of the patch transformers is 6.622 cm + 6.378 cm = 13.000 cm.

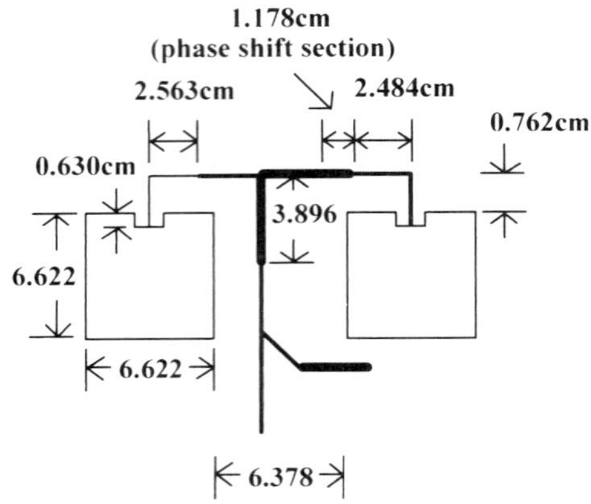

Figure 7.27 Feed layout with known dimensions.

Of this, 2.563 cm + 1.178 cm + 2.484 cm = 6.225 cm is accounted for. The remaining distance is 6.775 cm. One of the equations to be satisfied is therefore

$$l_l + l_r = 6.775 \tag{7.26}$$

The electrical lengths must also be equal. The wavelength in the 74-Ω line is 14.882 cm, which yields a phase shift of 360 degrees/14.882 cm = 24.19 deg/cm. The phase shift for the 154.1-Ω line is 23.38 deg/cm. The condition for equal electrical lengths is stated as

$$23.38[l_l - 0.0754] = 24.19[l_r - 0.0361] \tag{7.27}$$

where the lengths have been shortened by the reference plane extensions of the T-junction. Solving these two equations results in l_l = 3.464 cm and l_r = 3.322 cm.

The remaining length in the subarray to be found is the line that runs to the second splitter. The line has the same impedance and length for both halves of the subarray because of the in-phase, equal amplitude excitation. The 100-Ω lines from the patch pairs combine at the junction to produce a 50-Ω impedance. Since the output lines of this splitter are equal, so are the reference plane shifts. Determining the line lengths is simply a matter of finding the distance from the 70.7-Ω transformer to the T-junction. The vertical distance between the first level splitters is the same as the patch spacing, 13.000 cm. Part of this is taken up by the 70.7-Ω transformers. The length of the 100-Ω lines is [13.000 – 2(3.896)]/2 = 2.604 cm.

The final configuration for the subarray is given in Figure 7.28. Two 2.604-cm-long 100-Ω lines meet halfway between the horizontal feed line sections. At this point

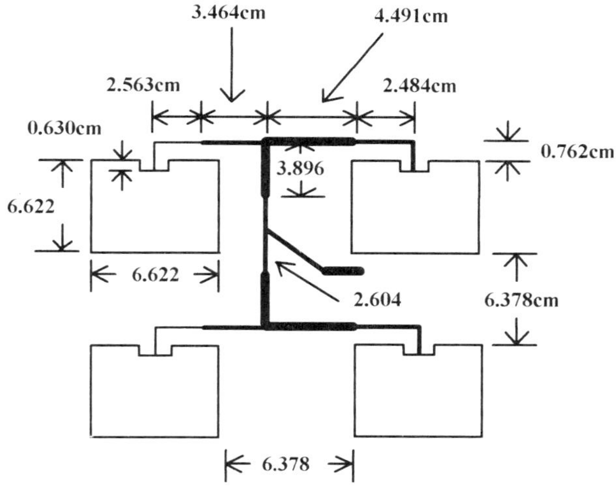

Figure 7.28 Final design of the subarray.

another transformer, whose value will be determined shortly, is located. The transformer is brought out from the junction at a 45-degree angle. The length of the angled section is such that when bent back to a horizontal orientation, the line is again 0.762 cm away from the patch edge.

The rest of the array design follows in a similar manner. The elements in the subarray adjacent to that just under consideration have amplitudes of 0.908 and 1.000. The power split ratio is 1.1013. The output line impedances are then $2.1013Z_0$ and $1.9080Z_0$. The splitter connecting the adjacent subarrays must provide a –57.0-degree phase shift (see Figure 7.23). It must also supply the proper amplitude ratio. Let P_{in} be the power entering the splitter connecting the leftmost two patches. This would be the power entering the 3.896-cm-long transformer of Figure 7.28. The power reaching the right patch of that pair is $(2.0823/3.0823)P_{in} = 0.676P_{in}$. Let P'_{in} be the power entering the splitter connecting the patches in the next subarray. The power delivered to the right patch of the pair is $(1.1013/2.1013)P'_{in} = 0.524P'_{in}$. The ratio of the powers in the two patches must be that of the desired amplitude distribution. Since the patch on the far right has a relative amplitude of 1 and that in the other pair of 0.658, the following relationship occurs between P_{in} and P'_{in}:

$$(0.676)P_{in} = 0.658(0.524)P'_{in} \tag{7.28}$$

which results in $P'_{in}/P_{in} = 1.959$. This is the power ratio for the splitter connecting the subarrays. The splitter output line impedances become $1.511Z_0$ and $2.959Z_0$. The latter is the impedance that the unknown final transformer in Figure 7.28 must match. Again using $Z_0 = 50\Omega$, the output line impedance is 148Ω, which requires a transformer impedance of 86.0Ω. When laying out this final transformer of the subarray, do not forget to increase its length by the reference plane shift in the T-junction.

One factor that has not been accounted for in the design are the effects of the step change in widths at the junctions of two lines with differing impedances. The parasitics associated with the width change will affect the phase and amplitude relationships and also the impedance match. If desired, the program MICSTEP may be used to find the junction-equivalent circuit elements. Calculating their effect requires some effort, which is why they are usually not included in the analysis. If microwave circuit design software such as TOUCHSTONE, COMPACT, or similar programs are available, they can be used to design the array. Since the programs can more rigorously account for, for example, parasitics and frequency variations, the reader is strongly urged to use them to assist in the design.

7.6 RESONANT SERIES-FED ARRAY DESIGN

With series-fed arrays, the radiating elements can either be in series or shunt across the feed line. In the microstrip arrays of Figure 7.15(a) the patches are in series, while in Figure 7.15(b) they are in shunt. The analysis given in this section will be applicable to both types of array. For broadside radiation, resonant arrays are used. The elements are spaced a wavelength apart where the wavelength is that of the transmission line. Let the line have a characteristic impedance of Z_0. Assume that at the design frequency, the radiating element impedance is purely real. An equivalent circuit for the series-fed array is easily constructed and is given in Figure 7.29. The circuit consists of shunt conductances, representing the elements, connected together by transmission line segments. The conductances shown in the figure are normalized with respect to the feed line admittance. Normalized values will be denoted with lowercase letters. Capital letters are used for the corresponding unnormalized quantities. With shunt elements, the array may be terminated in a short-circuited quarter-wavelength line, a half-wavelength open-circuited line, or just simply an open circuit immediately after the last element. The function of the termination

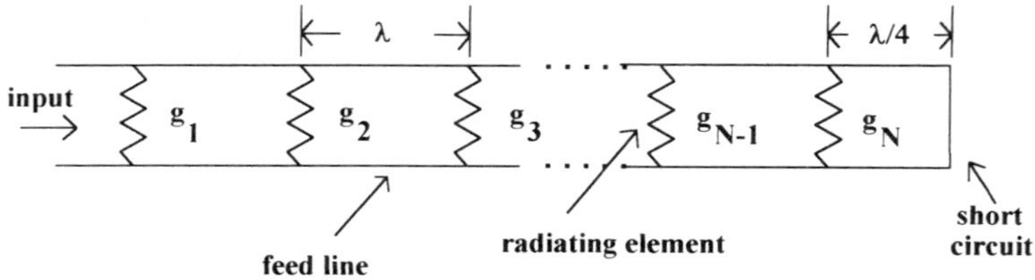

Figure 7.29 Series-fed array equivalent circuit.

is to ensure a voltage maximum at each element and to make any reflected signal add up in phase with the incident one. The analysis that follows is taken from [9].

Because of the wavelength spacing, the voltage across all the elements is the same. The power absorbed by the nth element is

$$P_{\mathrm{n}} = \frac{1}{2}V^2 g_n \tag{7.29}$$

where V is the voltage across each element. From (7.6), the far field radiated by the nth element is

$$E_n(\theta) = E_{\mathrm{e}}(\theta) a_n e^{j(n-1)kd\sin(\theta)} \frac{e^{jkr}}{r} \tag{7.30}$$

The power radiated by the element is proportional to the field squared, for example,

$$P_{\mathrm{r}n} \propto E_n^2 \propto a_n^2 \tag{7.31}$$

This power, in turn, is related to the amplitude excitation coefficient. The absorbed power (7.29) must equal the power radiated (7.31), ignoring the usually negligible losses in the antenna. From this conservation of power argument, the element conductance must be proportional to the amplitude coefficient squared.

$$g_n = K a_n^2 \tag{7.32}$$

where K is the constant of proportionality. Since the elements are spaced a wavelength apart, the input conductance to the array is the sum of all the element conductances

$$g_{\mathrm{in}} = \sum_{n=1}^{N} g_n \tag{7.33}$$

where N is the number of elements. Remembering that the conductances are normalized, the condition for a matched input is

$$g_{\mathrm{in}} = \sum_{n=1}^{N} g_n = 1 \tag{7.34}$$

K is found by combining (7.32) with (7.34) as

$$K = \frac{1}{\sum_{n=1}^{N} a_n^2} \tag{7.35}$$

With K known, it is possible to find the required element conductances from the amplitude distribution. The conductances can then be linked to some element of the design of the radiator. For example, with rectangular patches, the conductance is proportional to the patch width. This analysis ignores the external mutual coupling between elements caused by the near and far fields of the radiators. Except for low sidelobe designs, this is not a significant factor. The analysis also ignores feed line losses. This is a reasonable assumption as long as the power absorbed by the smallest conductance is larger than the line losses between elements. Some errors may be incurred for very long arrays.

The above derivation holds for an end-fed array. With a center-fed array, the expression for K is [10]

$$K = \frac{2}{\sum_{n=1}^{N} a_n^2} \tag{7.36}$$

where N is still the total number of elements, that is, the sum of both sides of the array.

As discussed earlier, the bandwidth of a resonant array is very narrow because the impedance match rapidly degrades with frequency. The reflections from each element cancel at the input only at the design frequency. An approximate rule of thumb for the bandwidth of a resonant array is [11]

$$BW \approx \frac{\pm 50\%}{N} \tag{7.37}$$

where the bandwidth criteria is an SWR $\leq$ 3:1.

The equivalent circuit of Figure 7.29 is directly applicable to the shunt-connected patch array of Figure 7.15(b). At resonance, the patch impedance is real. If the microstrip lines connecting the patches to the main feed line are a quarter-wavelength long, then the impedance seen by the main feed line is also real. The magnitude of the impedance and therefore the excitation is controlled by the characteristic impedance of the quarter-wavelength section. This line transforms the patch impedance into a lower value. The wider the line width, the lower the resulting shunt conductance and more power that is coupled. The flexibility of this approach is limited by the narrowest line that can be reasonably fabricated. High shunt conductances are often needed for elements near the input especially for larger arrays. Narrow main feed lines help since the normalized conductance is lowered but suffer from increased losses, which reduce antenna efficiency.

The circuit of Figure 7.29 is also valid for the vertical series feeds in Figures 7.18 and 7.19. Generally the rows of series-connected patches are designed to have an identical input impedance. The line connecting the rows to the vertical feed line are again quarterwave transformers to adjust the shunt impedance for the correct amplitude. The vertical line sections between rows are made a (microstrip) wavelength long to provide in-phase excitation.

The above analysis ignores junction effects, in particular the parasitic susceptance and reference plane extensions. The latter can be accounted for by adjusting the lengths of any line connected to the T-junction. Including the susceptance is more difficult. As with the corporate arrays, a microwave circuit design program should be used if more accuracy is needed.

The equivalent circuit of the series-connected patch array in Figure 7.15(a) is more involved. For one, the feed line is not homogeneous but consists of alternating sections. One is the narrow line connecting adjacent patches, and the other is the patch itself. Also the patch edges have a shunt capacitance associated with the fringing fields. The equivalent circuit is more like that shown in Figure 7.30. Each patch places two conductances and capacitors across the feed line. Since the patch is nominally a half-wavelength long, the line in between patches is also a half-wavelength long. Obviously, the characteristic impedance of the patch line is much lower than that for the interconnect.

The capacitors can be removed by replacing the physical length of the patch with an equivalent length that includes the end-effect extensions. This is actually the length calculated by programs like PATCH9. The voltages across the input- and output-edge conductances are equal but out of phase. Each edge absorbs essentially the same power as given by (7.29). The total power absorbed by the patch is therefore twice that of (7.29). Because of this, the equivalent circuit can be further simplified by combining the edge conductances as in Figure 7.31. The patches are now represented by a shunt conductance of twice the edge value. A fictitious line connects the patches. This circuit is now equivalent to Figure 7.29, so the same analysis may be used to determine the element conductances from the amplitude distribution. It must be emphasized that Figure 7.31 is an extremely simplified representation that should only be used to find the element conductances. The actual feed line length between patches is set at a half-wavelength. The patch lengths are determined by an analysis similar to that used in PATCH9.

The preceding analysis will be applied to the design of a series-connected patch array as in Figure 7.15(a). For series-fed arrays, the substrate choice will affect the patch size plus the array size. The length of both the patch and intervening feed line are direct functions of the dielectric constant. For a fixed number of elements an array built on a substrate with $\epsilon_r = 2.2$ will be larger than one built on an $\epsilon_r = 10$ substrate. This has

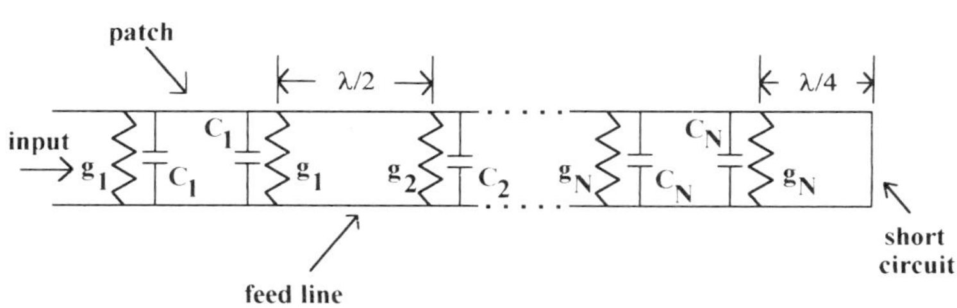

Figure 7.30 Equivalent circuit for a series-connected patch array.

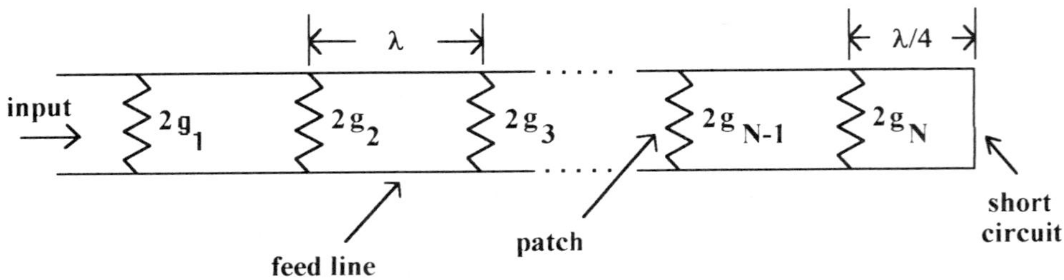

Figure 7.31 Simplified equivalent circuit for determining conductance values.

implications regarding the realized beamwidth and consequently the gain, as these parameters are dependent on the antenna size measured in terms of free-space wavelengths. Surface-wave excitation is lower with smaller dielectric constants. As a rule, lower ϵ_r materials are generally preferred.

The first step is to perform an initial sizing of the array. To do this, some estimate of the patch size and feed line dimensions is needed. Assume an array is to be built to operate at 1.5 GHz. The substrate is chosen to be the same as in the corporate array design example, that is $\epsilon_r = 2.2$, tan $\delta = 0.001$, height = 0.160 cm, and conductor thickness = 0.00356 cm. Again, a square patch provides a representative example of the patch size to expect. From PATCH9 (run with feed type of none and no inset), the patch dimensions are 6.622 cm × 6.622 cm.

There are several factors to consider when choosing the feed line width. In order for the patches to operate like patch antennas, the feed line width should be a small fraction of the antenna width. Otherwise, significant blockage of the radiating edges occurs. With end-fed arrays, the elements nearest the feed couple only a small amount of power and therefore must be fairly narrow. The feed line must be small compared to the narrowest patch. When establishing the coupling to each element, it is the element impedance relative to the feed line's that matters. This is controlled by the patch width. The narrowest usable width is determined by the feed line width. The widest is determined by the need to keep the patch less than a wavelength wide. The widest range of coupling values occurs when the feed line impedance is high. On the other hand, extremely narrow lines suffer from high losses. Some compromise is usually necessary. In this case, a 90-Ω line will be used. This is a 0.1737-cm-wide line that is considerably smaller than the square patch width.

The effective dielectric constant of the feed line is $\epsilon_{re} = 1.771$ making a half-wavelength section 7.582-cm-long. The element spacing (that is the distance between input edges of adjacent patches) is 6.622 cm + 7.582 cm = 14.204 cm. The length of an array of x elements can now be estimated. The array consists of x elements plus $x - 1$ feed line sections (the input feed line is not counted as part of the radiating array). For example, a ten-element array will be approximately

10(6.622 cm) + 9(7.582 cm) = 134.458-cm long. ARRAYCAL and APERDIST can be used to determine the array size and amplitude distribution needed to meet the design requirements. Assume that it is found that a ten-element array with a linear taper having a 6-dB pedestal will be required. The amplitude distribution as determined by APERDIST is as follows.

APERDIST.V12   01-07-1996   16:03:18

LINEAR ARRAY EXCITATION COEFFICIENTS

FREQUENCY = 1.500 GHz
ARRAY LENGTH = 127.833 cm
NUMBER OF ELEMENTS = 10
APERTURE DISTRIBUTION = LINEAR

ELEMENT	POSITION (cm)	AMPLITUDE (deg)	PHASE
1	–63.916	0.501	0.00
2	–49.713	0.626	0.00
3	–35.509	0.751	0.00
4	–21.306	0.875	0.00
5	–7.102	1.000	0.00
6	7.102	1.000	0.00
7	21.306	0.875	0.00
8	35.509	0.751	0.00
9	49.713	0.626	0.00
10	63.916	0.501	0.00

(The array length shown in the output is slightly different from the physical length of 134.458 cm. In the program, the electrical length of the array is considered to be from the center-to-center spacing between edge elements, not the outside edge to outside edge.) The next step is to find the element conductances needed to realize this distribution.

The element conductances are determined using (7.32) to (7.36). These equations have been programmed in RESFEED, which analyzes either end-fed or center-fed resonant arrays. RESFEED converts a desired amplitude distribution into the required element conductances. Running the program for this example yields the following output.

RESFEED.V11   9/25/95   10:15:05
[Current Program Date - 12/4/95]

Calculates conductances for resonant array of shunt elements

CENTER (c) OR END (e) ARRAY? **e**

INPUT TOTAL NUMBER OF ELEMENTS IN ARRAY? **10**

INPUT ARRAY EXCITATION FROM KEYBOARD (k) OR FILE (f)? **k**

(The amplitude distribution can be read in from a file. The file format is identical to that used by NARRAYD. There is one line per element. Each line contains the element position (cm), excitation amplitude, and phase (deg) separated by a commas. If the file option is chosen, RESFEED asks for the file name.)

The following inputs are skipped if the file option is used.

UNIFORM (u) OR TAPERED (t) ARRAY EXCITATION? **t**

(For the uniform distribution, there is no need to read in the excitations.)

INPUT ARRAY EXCITATION
[Start with first element for either type of array]

(In other words, for either array, start with the element that is closest to the input. For a center-fed array, only input for half the array starting at the feed and going to the end.)

INPUT EXCITATION FOR ELEMENT NUMBER 1? **0.501**

INPUT EXCITATION FOR ELEMENT NUMBER 2? **0.626**

INPUT EXCITATION FOR ELEMENT NUMBER 3? **0.751**

.

.

.

This continues (following the excitations calculated by APERDIST) until all elements have been input. The final input is

INPUT NORMALIZING IMPEDANCE (ohms)? **90**

RESFEED then calculates the element conductances and displays the results on the screen as follows.

RESFEED.V11 09-26-1995 22:19:03

Shunt Conductances for a Series Fed Array

END FED ARRAY
NUMBER OF ELEMENTS = 10
NORMALIZING IMPEDANCE = 100.00
Gsum = 1.000

ELEMENT	EXCITATION	NORMALIZED g	ACTUAL g
1	0.501	0.042220	0.000422
2	0.626	0.065917	0.000659
3	0.751	0.094870	0.000949
4	0.875	0.128785	0.001288
5	1.000	0.168208	0.001682
6	1.000	0.168208	0.001682
7	0.875	0.128785	0.001288
8	0.751	0.094870	0.000949
9	0.626	0.065917	0.000659
10	0.501	0.042220	0.000422

Note that both the normalized and unnormalized values are presented. Options to print and store the data are available.

The last step in the design procedure is to determine the patch widths needed to realize the conductances. This information is contained in the transmission line model since it includes the radiating edge conductance. PATCHCOM is a modified version of PATCH9. Like PATCH9, it determines the resonant length of the patch. Unlike PATCH9, PATCHCOM accounts for the presence of the feed line at both edges to simulate the array situation. Also PATCHCOM only calculates the impedance at the design frequency. It does, however, calculate the patch parameters over a range of widths. This makes it useful for finding the widths needed to establish the amplitude distribution.

Virtually all of the inputs to the program are identical to PATCH9 (there is only a design mode), so will not be repeated here. The only substantial difference is that no frequency range for impedance calculation is requested. Instead, the program asks

Input minimum and maximum patch widths (cm)? **1.778,4.572**

(It may take several runs to get all the widths needed, especially if the conductance values span a large range.)

Input number of patch widths to analyze? **22**

At this point, the program calculates each width to be used. If a width exceeds a wavelength, the program prints a warning flag and asks if the user wants to change that particular value. As in PATCH9, the final input is

Input fractional change (%) in patch length for initial search?
(default = 1%)

0.1

(This was put in PATCH9 to help in situations such as thick or very thin substrates where the impedance changes with length are dramatic and the normal length increment is too large. With PATCHCOM, convergence difficulties may be experienced with narrow patch widths. It is strongly recommended that a fractional change of 1% or less be routinely used to avoid convergence problems.)

For each width, the program determines the resonant length and input conductance at resonance. The edge conductance and capacitance is modified by the feed line width on both ends, so the results differ slightly from those predicted by PATCH9. The output in this case is as follows.

PATCHCOM.V23 01-07-1996 13:40:27
Current Program Date - 01/07/1996

Substrate height (cm) = 0.160
Relative dielectric constant = 2.200
Loss tangent = 0.0010
Line thickness (cm) = 0.004
Feed type = microstrip line
Feed line width (cm) = 0.17
Resonant frequency (GHz) = 1.500000

WIDTH	SLOT G (cm)	PATCH Zo (mhos)	PATCH er (ohms)	PATCH LENGTH (cm)
1.778	0.000360	18.80	2.03	6.902
1.905	0.000400	17.73	2.03	6.884
2.032	0.000442	16.77	2.04	6.868
2.159	0.000486	15.92	2.05	6.853
2.286	0.000533	15.15	2.05	6.839
2.413	0.000581	14.45	2.06	6.827
2.540	0.000631	13.81	2.06	6.815
2.667	0.000683	13.23	2.07	6.804
2.794	0.000738	12.69	2.07	6.794
2.921	0.000794	12.20	2.08	6.785
3.048	0.000852	11.75	2.08	6.776
3.175	0.000912	11.32	2.09	6.768
3.302	0.000973	10.93	2.09	6.760
3.429	0.001037	10.56	2.09	6.753
3.556	0.001103	10.22	2.10	6.746
3.683	0.001170	9.90	2.10	6.740
3.810	0.001239	9.60	2.10	6.734
3.937	0.001310	9.32	2.10	6.728
4.064	0.001383	9.05	2.11	6.722
4.191	0.001457	8.80	2.11	6.717
4.318	0.001534	8.56	2.11	6.712
4.445	0.001611	8.34	2.11	6.708
4.572	0.001691	8.12	2.12	6.703

Note that this covers the range of conductances determined by RESFEED. As usual, the data may be printed out or stored in a file.

The patches needed for the array are selected from the above output and the required conductances. They are listed in Table 7.2. The patch lengths for each width are also output by PATCHCOM. With the patch dimension and interconnect length known, the design is virtually complete.

A sketch of the array is given in Figure 7.32. At the output (right) end, the array can be left as is with no termination. Another option is to use another half-wavelength section of feed line that is open-circuited. The array can also be terminated with a short-circuited quarter-wavelength line. The final part to consider is the input line. The array impedance will be 90Ω at the design frequency. A quarterwave transformer may be inserted to convert this to a more appropriate value.

If this array is part of a larger two dimensional array such as those of Figures 7.18 and 7.19, then essentially the same design procedure is used. The vertical feed line does

Table 7.2
Patch Widths to Realize Linear Taper
With a 6-dB Pedestal

Conductance (mhos)	*Patch Width (cm)*
0.000442	1.968
0.000659	2.604
0.000949	3.238
0.001288	3.874
0.001682	4.572
0.001682	4.572
0.001288	3.874
0.000949	3.238
0.000659	2.604
0.000442	1.968

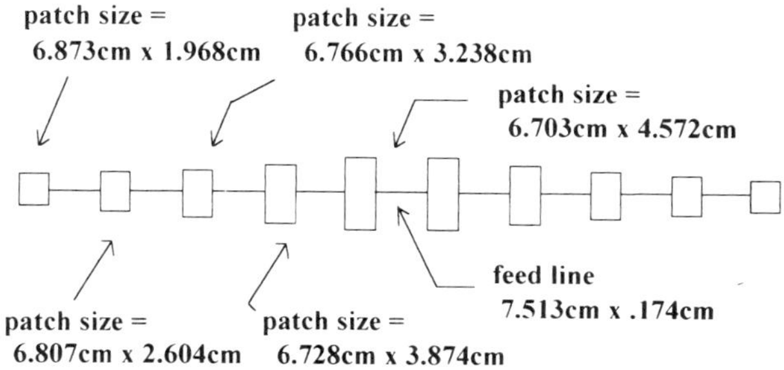

Figure 7.32 Ten-element resonant series-fed array.

not have to be as high in impedance as for the array just designed. There is no concern about blockage of the radiating element. The impedance just needs to be high enough to allow sufficient flexibility in the coupling. The coupling is determined by the impedance presented by the row array to the vertical feed. This is controlled by the quarterwave transformer at the input to each array. The feed line length between rows of arrays is equal to a wavelength in the feed line. This distance should be corrected for any reference plane shifts in the T- or cross-junctions.

The design of the series-fed array of Figure 7.15(b) follows along similar lines. The required conductances are found as before. ARRAYCAL and APERDIST are used to determine the amplitude distribution, and RESFEED converts the distribution into conductance values. The conductance across the feed line due to a patch is a function of the widths of the patch and the line connecting the patch to the main feed. The latter is a quarterwave transformer that takes the patch impedance and brings it down to a lower value. Since typical conductances are small, it is important to keep the main feed line, branch line, and patch impedances as high as possible. With the main feed line there is always the trade-off between high impedance and high line losses.

Generally it is advisable to keep the patches identical along the array. The patch width is determined by the conductance values and the widths to be used for the branch lines. The smallest conductance in the array places the most restriction on the patch and branch line widths. The patch width should be chosen such that the branch line width needed to transform to the required conductance is at least as wide as the minimum that can be reasonably fabricated. This places a lower limit on the patch width. On the opposite end, the branch line width needed for the highest conductance should not become a significant fraction of the patch width. If both criteria cannot be simultaneously met, it may be necessary to select a different main feed line width or perhaps consider a different amplitude distribution. The use of inset fed patches is perhaps another option.

7.7 SERIES-FED TRAVELING WAVE ARRAY DESIGN

When the main beam must be positioned off of broadside, series-fed traveling wave arrays are one option. In this case, the elements are spaced at a distance determined by the desired beam angle. Consider the array of Figure 7.15(b). If all the branch lines are a quarter wavelength long, then the element-to-element phase shift is a function of the length of the main feed between patches. This phase shift is given by $\beta = k_m d$. k_m is the phase constant of the microstrip line, $k_m = 2\pi\sqrt{\epsilon_{re}}/\lambda_0$. d is the length of the feed line. Although shown straight in the figure, the feed line can be meandered between patches to increase its length. If the line is straight, d is also the element spacing. To position the beam at the angle, θ_0, the following equation must be satisfied

$$(k_0 \sin(\theta_0) - k_m)d = 2m\pi \tag{7.38}$$

where a straight feed line is assumed and $m = 0, \pm 1, \pm 2, \ldots$. Given the operating frequency, the microstrip feed line effective dielectric constant, and m, (7.38) can be solved for the line length via

$$d = \frac{m\lambda_0}{\sin(\theta_0) - \sqrt{\epsilon_{\text{re}}}} \tag{7.39}$$

where use has been made of the relationship, $k_m = \sqrt{\epsilon_{\text{re}}}k_0$. Note that m must be chosen such that d is positive.

The situation for the array in Figure 7.15(a) is somewhat more complicated. Here the element-to-element phase shift contains two components. One is the phase shift through the patch. This is slightly less than 180 degrees. The other is the shift through the interconnecting feed line. Let k_{p} be the phase constant for the patch, l_{p} be the patch length, and l_{f} be the length of the feed line. The equation to be satisfied is

$$k_0 d \sin(\theta_0) - [k_{\text{p}} l_{\text{p}} + k_{\text{m}} l_{\text{f}}] = 2m\pi \tag{7.40}$$

Again, this can be solved for the feed line length. For a straight feed line, $d = l_{\text{p}} + l_{\text{f}}$, so the solution for the feed line length becomes

$$l_{\text{f}} = \frac{m\lambda_0 - [\sin(\theta_0) - \sqrt{\epsilon_{\text{rep}}}]l_{\text{p}}}{\sin(\theta_0) - \sqrt{\epsilon_{\text{re}}}} \tag{7.41}$$

where ϵ_{rep} is the effective dielectric constant of the patch.

For both types of arrays, the smallest value of m is usually used. This results in less beam steer with frequency and avoids the appearance of multiple main beams. It is advisable to find the phase shift that results from the solution of (7.39) or (7.41) and to calculate the pattern using NARRAYD (for (7.41) add the additional phase shift of the patch). This will show if spurious main beams will occur because of the element spacings.

The analysis for the conductance values follows a similar approach to that of the resonant array [12]. There are a few differences, however. Each conductance is assumed small enough that any reflection is negligible. Also the array is terminated in a matched load. There is essentially no reflected wave present, and this has two implications. One is that the array is matched at the input over a band of frequencies. With the element spacing not being an integral number of half-wavelengths and the conductances small, any residual reflections add up at the feed in an incoherent manner. The second is that with little reflected signal, the power coupled by an element is found from the incident power only. This simplifies the analysis considerably. The maximum normalized conductance should be no more than 0.2 for these assumptions to hold. Unfortunately to maintain this, the analysis is strictly valid only for arrays of about 10 elements or more.

When the conductances are constrained as above, it is not possible to radiate all of the power input to the array. A certain amount, P_{l}, is left after the last element. Even with no feed line losses, the efficiency of a traveling wave array is less than 100%. This is the trade-off for realizing wider bandwidth operation with the traveling wave array. The termination must be a matched load to absorb this residual power. Any reflection from the termination shows up as a main beam on the opposite side of the array normal to the desired beam angle. The magnitude of this second beam depends upon the amount of reflected power. In most cases, the pattern of the reflected beam increases the overall sidelobe level. For low sidelobe designs, it is important to minimize the reflected signal.

The magnitude of P_l is a parameter under the control of the designer. It should be chosen low enough to maintain adequate efficiency. P_l is defined as the amount of the input power dissipated in the load. The remainder is radiated. The efficiency is therefore immediately specified by the load power. For example, if P_l is -10 dB, then the efficiency is 90%. There is a minimum value of P_l for any design that depends upon the maximum conductance that can be realized. For example, in the array of Figure 7.15(b), the patch and branch line widths have restrictions as mentioned earlier. These limit the conductance that can be placed across the feed. Also too low a value of P_l requires normalized conductances, which may exceed 0.2 and violate the criteria for low reflections. The design of the array sometimes requires several iterations with different values of P_l in order to obtain a suitable conductance distribution.

To determine an expression for the conductances, start at the last element [12]. Let the power absorbed by this element be P_N. From (7.32) this is proportional to the square of the required excitation amplitude, a_N. Since the power left after the last element is the load power, P_l, the power incident must be $P_l + P_N$. The normalized conductance is therefore

$$g_N = \frac{P_N}{P_l + P_N} \tag{7.42}$$

The power incident on the next to last element is $P_l + P_{N-1} + P_N$. The conductance is

$$g_{N-1} = \frac{P_{N-1}}{P_l + P_{N-1} + P_N} \tag{7.43}$$

Continuing with this analysis, an expression can be found for any conductance in the array

$$g_n = \frac{P_n}{P_l + \sum_{i=n}^{N} P_i} = \frac{P_n}{1 - \sum_{i=1}^{n-1} P_i} \tag{7.44}$$

where the second expression results from the fact that the sum of the radiated and load powers is set equal to one. This allows a determination of the constant of proportionality, K, between a_n and P_n. Since the total power is equal to one,

$$1 = r + \sum_{n=1}^{N} P_n = r + \sum_{n=1}^{N} Ka_n^2 \tag{7.45}$$

Solving for K

$$K = \frac{1 - r}{\sum_{n=1}^{N} a_n^2} \tag{7.46}$$

With this normalization of the powers,

$$g_n = \frac{Ka_n^2}{1 - \sum_{n=1}^{N} Ka_n^2} \tag{7.47}$$

which is analogous to (7.32). Keep in mind that the g_n of (7.47) is normalized.

To design a traveling wave array like Figure 7.15(a), start by finding the size of a square patch as was done for the resonant array. Choose a feed line width using the same considerations as before; that is, it should be much narrower than the smallest patch width. Find the effective dielectric constants of the patch and the feed using MICRO. Choose a value for m and, using (7.41) along with the desired beam angle, calculate the feed line length. This gives an approximate element spacing. Use ARRAYCAL and APERDIST to find the number of elements and distribution based upon the pattern requirements. When investigating aperture distributions with APERDIST, checks for spurious main beams can also be done. One of the inputs is the element-to-element phase shift, which is given by $k_p l_p + k_m l_f = (360/\lambda_0)[\sqrt{\epsilon_{rep}} l_p + \sqrt{\epsilon_{re}} l_f]$. The program calculates the pattern to allow the user to determine whether the distribution is acceptable. While doing this, the presence of spurious beams can be detected.

APERDIST may also be used to ascertain the beam steer with frequency. Since the change in phase shift is linear with frequency, if β_0 is the phase shift at the design frequency, f_0, then at frequency f, it is

$$\beta = \frac{f}{f_0}\beta_0 \tag{7.48}$$

As long as the frequency is not too far from f_0, the amplitude distribution can be considered unchanged from the design value. Try calculating patterns at different frequencies by using the phase shifts found from (7.48). The pattern calculations help determine if the selected value for m is appropriate. If not, another iteration may be necessary. Typically, a wrong value of m results in a feed line length that is too long. The element spacing becomes too large, which causes spurious main beams and rapid beam steer with frequency.

As an example, consider an array at 1.5 GHz with a main beam at 5 degrees off broadside. Assume the usual 0.160-cm-thick substrate with $\epsilon_r = 2.2$. Since we need to know the patch effective dielectric constant, PATCHCOM is run to determine the initial dimensions. For these calculations, a feed line width of zero is used. After several inputs have been requested, PATCHCOM asks for a patch length giving 6.742 cm as an estimate. Input this value and then input a range of widths that straddle the length, since the actual length typically comes out less than the estimate. In this case, a range from 5.080 cm to 7.62 cm is used. The output shows that the patch length is about 6.642 cm or less over this range of inputs. After looking at the results, it appears that a square patch lies somewhere between a width of 6.629 cm and 6.655 cm. The program is run again between this range, and from the output it can be determined that the appropriate dimension is 6.640 cm. This number differs by 0.0178 cm from that previously found using PATCH9 because of slight differences in the program to include the effect of a feed line at both edges.

The feed line must be narrow enough to accommodate the various patch widths. Choose a 100-Ω feed line. This requires a 0.138-cm-wide line (from MICRO). The feed effective dielectric constant is 1.753. The value of m must be less than zero for the feed line length to come out positive. Remembering that a low value is preferred, m is taken to be −1. Substituting into (7.41), the result is $l_f = 8.786$ cm. The approximate element spacing is 6.640 cm + 8.786 cm = 15.426 cm.

Assume that after using ARRAYCAL and APERDIST, a 10-element array with a linear taper having a 6-dB pedestal is adequate for the pattern specifications. The phase shift between elements is −384.18 degrees. The pattern, as calculated by APERDIST, shows the beam peak at 5 degrees with no evidence of secondary peaks. Therefore, $m = -1$ is a good choice for this antenna. Arrays like those in Figure 7.15 are prone to forming additional beam peaks because, with a straight feed line, the element spacing becomes quite large. This limits the amount of beam steer that can be realized. One way around this is to meander the feed line between patches as in Figure 7.33. Both types of arrays in Figure 7.15 can be configured this way. By placing the patches closer together, the onset of grating lobe formulation is pushed further out allowing larger beam angles. Unfortunately this approach increases the beam steer with frequency.

With meandered feed lines, (7.39) and (7.41) become

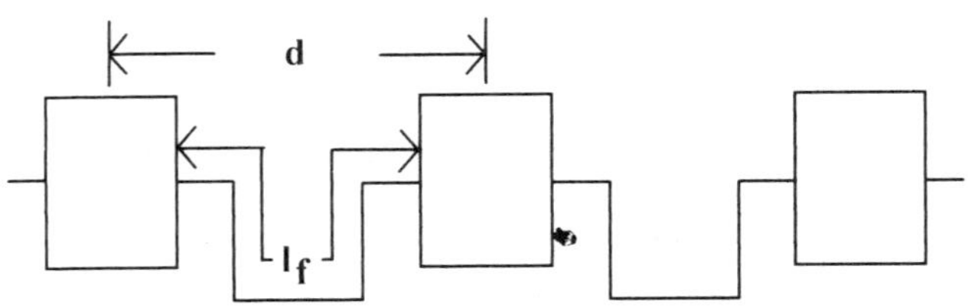

Figure 7.33 Series-fed array with meandered feed lines for increased beam steer.

$$l_f = \frac{d\,\sin(\theta_0) - m\lambda_0}{\sqrt{\epsilon_{re}}} \tag{7.49}$$

and

$$l_f = \frac{d\,\sin(\theta_0) - \sqrt{\epsilon_{rep}}\,l_p - m\lambda_0}{\sqrt{\epsilon_{re}}} \tag{7.50}$$

respectively.

Returning to the array design, the amplitude distribution as determined by APERDIST is as follows.

APERDIST.V10 10-01-1995 18:34:36

LINEAR ARRAY EXCITATION COEFFICIENTS

FREQUENCY = 1.500 GHz
ARRAY LENGTH = 145.465 cm
NUMBER OF ELEMENTS = 10
APERTURE DISTRIBUTION = LINEAR

ELEMENT	POSITION (cm)	AMPLITUDE (deg)	PHASE
1	−72.733	0.501	0.00
2	−56.571	0.626	−384.18
3	−40.406	0.751	−768.37
4	−24.244	0.875	−1152.55
5	−8.082	1.000	−1536.73
6	8.082	1.000	−1920.92
7	24.244	0.875	−2305.10
8	40.406	0.751	−2689.28
9	56.571	0.626	−3073.46
10	72.733	0.501	−3457.65

This, of course, is virtually identical to the resonant array example. The element conductances must be found next. Program CONDCAL does this by evaluating (7.47).

PROGRAM CONDCAL.V10 09/29/95 13:33:23
[Current Program Date - 25/09/95]

INPUT NUMBER OF ELEMENTS? **10**

INPUT ELEMENT SPACING? **15.425**

INPUT FEED LINE CHARACTERISTIC IMPEDANCE (ohms)? **100**

INPUT FEED LINE ATTENUATION (dB/cm)? **0**

(CONDCAL can account for attenuation, which modifies the conductances, in the feed line. For the array of Figure 7.15(a), the attenuation is a combination of that due to the patch and the feed line. MICRO will find these attenuations. Each should be multiplied by the respective line length, added together, and then averaged by dividing by the sum of the lengths. With Figure 7.15(b) only the attenuation of the feed line need be considered. For this example, the attenuation is assumed negligible.)

INPUT EXCITATIONS (i) OR READ FROM FILE (r)? **i**

(This is identical to RESFEED. If a file is used, it must have one line of data per element. The line has three numbers, separated by commas, representing the element position (cm), amplitude, and phase (deg).)

INPUT EXCITATION FOR ELEMENT NUMBER 1? **0.501**

INPUT EXCITATION FOR ELEMENT NUMBER 2? **0.626**

.

.

.

This continues until all excitations are entered.

INPUT POWER INTO THE LOAD (negative dB)? **–10**

(This is a typical value to use for the amount of power left for the termination to absorb. Higher values (for example, –5 dB) degrade antenna efficiency and lower values (for example, –15 dB or so) are acceptable as long as the resulting conductances do not become too large. Lower values are needed for low sidelobe designs.)

The conductances are calculated and displayed as follows.

PROGRAM CONDCAL.V10   10-01-1995   18:37:19

NUMBER OF ELEMENTS = 10
ELEMENT SPACING = 15.425 cm
FEED LINE Zo = 100.00 ohms
FEED LINE ATTENUATION = 0.000 dB/cm
POWER IN LOAD = –10.00 dB
EFFICIENCY = 90.00%

ELEMENT	NORMALIZED CONDUCTANCE	CONDUCTANCE
1	0.0380	0.00038
2	0.0617	0.00062
3	0.0946	0.00095
4	0.1418	0.00142
5	0.2158	0.00216
6	0.2753	0.00275
7	0.2908	0.00291
8	0.3020	0.00302
9	0.3006	0.00301
10	0.2754	0.00275

The program calculates an approximate value for the antenna efficiency. This helps to assess whether the choice of P_l is acceptable or not. The choice for P_l appears to be reasonable. The resulting conductances are easily realized. Note that for the traveling wave array, the distribution is not symmetrical. It increases along the array until the last element (or elements for larger arrays) where it drops off again. This is typical for these types of arrays.

Also notice that the results violate the rule of thumb that the normalized conductances should be not more than 0.2. This is a difficulty with designing a traveling wave array with a relatively small number of elements. The only ways to reduce the maximum conductance are to either increase the power in the load, increase the number of elements, or perhaps change the aperture distribution. In this example, the maximum value can be reduced to 0.2 by increasing the load power to −6 dB. The efficiency drops to 74.9%. If this is acceptable, then this load power could be used. Since this is only an example, the design procedure will continue with the present −10 dB value. If the current conductances were actually used, the antenna would suffer from pattern distortions and a degraded match because of the reflections.

The final step is to use PATCHCOM to find the needed patch widths.

```
PATCHCOM.V23     01-07-1996        13:46:37
      Current Program Date - 01/07/1996

        Substrate height (cm) = 0.160
     Relative dielectric constant = 2.200
          Loss tangent = 0.0010
        Line thickness (cm) = 0.004
Feed type = microstrip line Feed line width (cm) = 0.14
      Resonant frequency (GHz) = 1.500000
```

WIDTH	SLOT G (cm)	PATCH Zo (mhos)	PATCH er (ohms)	PATCH LENGTH (cm)
1.778	0.000367	18.80	2.03	6.897
1.905	0.000408	17.73	2.03	6.879
2.032	0.000450	16.77	2.04	6.863
2.159	0.000495	15.92	2.05	6.849
2.286	0.000542	15.15	2.05	6.835
2.413	0.000590	14.45	2.06	6.823
2.540	0.000641	13.81	2.06	6.811
2.667	0.000694	13.23	2.07	6.800
2.794	0.000749	12.69	2.07	6.790
2.921	0.000805	12.20	2.08	6.781
3.048	0.000864	11.75	2.08	6.772
3.175	0.000924	11.32	2.09	6.764
3.302	0.000987	10.93	2.09	6.757
3.429	0.001051	10.56	2.09	6.750
3.556	0.001117	10.22	2.10	6.743
3.683	0.001185	9.90	2.10	6.737
3.810	0.001254	9.60	2.10	6.731
3.937	0.001326	9.32	2.10	6.725
4.064	0.001399	9.05	2.11	6.720
4.191	0.001474	8.80	2.11	6.714
4.318	0.001551	8.56	2.11	6.710
4.445	0.001629	8.34	2.11	6.705
4.572	0.001709	8.12	2.12	6.701
4.699	0.001791	7.92	2.12	6.697
4.826	0.001875	7.73	2.12	6.693
4.953	0.001960	7.54	2.12	6.689
5.080	0.002046	7.37	2.12	6.685
5.207	0.002134	7.20	2.13	6.682
5.334	0.002224	7.04	2.13	6.678
5.461	0.002316	6.89	2.13	6.675
5.588	0.002408	6.74	2.13	6.672
5.715	0.002503	6.60	2.13	6.669
5.842	0.002598	6.47	2.13	6.666
5.969	0.002695	6.34	2.14	6.664
6.096	0.002794	6.21	2.14	6.661
6.223	0.002894	6.10	2.14	6.659
6.350	0.002995	5.98	2.14	6.656
6.477	0.003097	5.87	2.14	6.654
6.604	0.003201	5.76	2.14	6.652
6.731	0.003306	5.66	2.14	6.650
6.858	0.003412	5.56	2.14	6.648

Using these results, the patch widths can be determined, interpolating for some of the values. These are shown in Table 7.3. The narrowest patch is still significantly wider than the feed line, which verifies the selection of a 100-Ω line. The final array configuration is shown in Figure 7.34. The patch dimensions are not shown but can be found in Table 7.3 and in the output from CONDCAL. The interconnecting feed line is 0.138-cm wide and 8.786-cm long.

After the last patch is another section of 100-Ω line to which is attached a 100-Ω resistor. The resistor is grounded on its opposite end to form the termination just as discussed in conjunction with Figure 7.16. A pad with a via pin to the ground plane may be used for this purpose. Any bond wires used to connect the resistor to the line and the pad should be kept as short as possible. At higher frequencies, the ground may be realized using a quarter-wavelength open-circuited line. Also at higher frequencies, the use of microwave chip resistors is recommended. They can be directly soldered to pads, avoiding the need to use bond wires. The input to the array has an impedance of 100Ω. A quarterwave transformer converts this to 50Ω or some other suitable value.

The design of shunt-connected patches (Figure 7.15(b)) follows a similar procedure. Use (7.39) to find the feed line length. Choose a small value for m, like −1. The phase

Table 7.3
Patch Widths for Traveling Wave Array
With 6-dB Taper

Conductance (mhos)	*Patch Width (cm)*
0.00038	1.778
0.00062	2.476
0.00095	3.238
0.00142	4.128
0.00216	5.207
0.00275	6.032
0.00291	6.223
0.00302	6.350
0.00301	6.350
0.00275	6.032

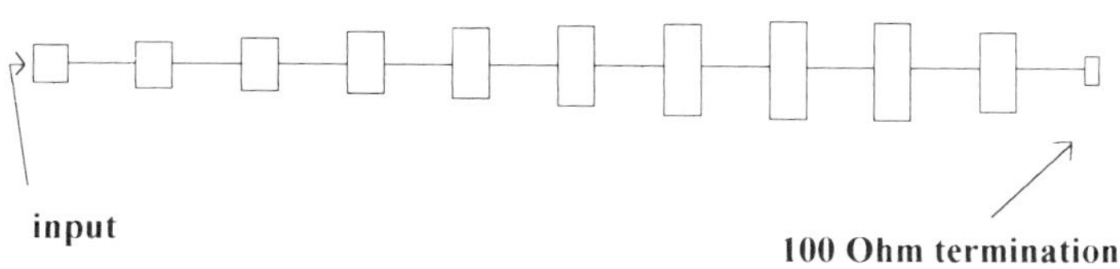

Figure 7.34 Traveling wave array design.

shift is $k_m l_f$. Use ARRAYCAL and APERDIST to determine the number of elements and distribution and then find the conductances with CONDCAL. Start with a reasonably high impedance patch, say a square one. Find the transformer widths needed to realize the conductances. If any width is too narrow, try a narrower patch width. Hopefully once the smallest conductance is obtained, the transformer width needed for the largest is not too wide. If it is a matter of only one or two conductances, either accept the wide feed or perhaps widen only the affected patches, assuming the width increase is not dramatic. Complete the design by adding the termination and any input transformer.

This same approach is used for the vertical feed sections in Figures 7.18 and 7.19. The conductances are obtained by adjusting the transformers at the inputs of the arrays connected to the vertical feed line.

As with the corporate-fed arrays, the analysis of either resonant or traveling wave arrays can be performed using a microwave circuit simulator. CONDCAL gives the input patch conductance at resonance. The conductance of an edge is simply half this value. Program MICENDEF determines the end-effect extension, Δl, by using the patch width for the line width. The edge capacitance is then given by

$$C_s = \frac{\tan(k_m \Delta l)}{2\pi f} \tag{7.51}$$

where f is the design frequency. The circuit program typically has built-in models for the microstrip line impedance and effective dielectric constant. When using a simulator, adjust the pertinent parameters (for example, patch width and main feed line length) until the amplitude and phase between adjacent elements are the desired values. For the most accurate results, the use of a circuit program is highly recommended.

References

[1] Elliott, R. S., ''Beamwidth and Directivity of Large Scanning Arrays, Pt. 1,'' *Microwave J.*, Vol. 6, No. 12, Dec. 1963, pp. 53–60.

[2] Elliott, R. S., ''Beamwidth and Directivity of Large Scanning Arrays, Pt. 2,'' *Microwave J.*, Vol. 7, No. 1, Jan. 1964, pp. 74–82.

[3] Stutzman, W. L., and G. A. Thiele, *Antenna Theory and Design*, New York, NY: John Wiley & Sons, 1981, Chap. 10.

[4] Balanis, C. A., *Antenna Theory Analysis and Design*, New York, NY: Harper & Row, 1982, Chap. 6.

[5] Rudge, A. W., K. Milne, A. D. Olver, and P. Knight, *The Handbook of Antenna Design*, Vol. 2, London, UK: Peter Peregrinus, Ltd., 1983, pp. 73–79.

[6] Ashkenazy, J., P. Perlmutter, and D. Treves, ''A Modular Approach for the Design of Microstrip Array Antennas,'' *IEEE Trans. on Antennas and Propagation*, Vol. 31, No. 1, Jan. 1983, pp. 190–193.

[7] James, J. R., P. S. Hall, and C. Wood, *Microstrip Antenna Theory and Design*, London, UK: Peter Peregrinus, Ltd., 1981, Chaps. 5 and 6.

[8] Jedlicka, J. P., M. T. Poe, and K. R. Carver, ''Measured Mutual Coupling Between Microstrip Antennas,'' *IEEE Trans. on Antennas and Propagation*, Vol. 29, No. 1, Jan. 1981, pp. 147–149.

[9] Collin, R. E., *Antennas and Radiowave Propagation*, New York, NY: McGraw-Hill, 1985, pp. 266–268.

[10] Johnson, R. C., and H. Jasik, *Antenna Engineering Handbook*, 2nd ed., New York, NY: McGraw-Hill, 1984, pp. 9-20–9-21.
[11] Hansen, R. C., *Microwave Scanning Antennas Volume III Array Systems*, New York, NY: Academic Press, 1966, p. 33.
[12] Collin, R. E., and F. J. Zucker, *Antenna Theory, Pt.* II, New York, NY: McGraw-Hill, 1969, pp. 595–598.

Appendix A
Microstrip Transmission Line and Discontinuity Programs

The design of microstrip antennas very often requires the calculation of microstrip line characteristics and the equivalent circuit elements for various microstrip discontinuities. Several programs have been written to provide this information to the designer. These are discussed in this appendix.

Program MICRO calculates the characteristic impedance, effective dielectric constant, and attenuation for a microstrip transmission line. It uses (2.1) to (2.8), (2.11) to (2.16) plus the corrections for dispersion presented in [1] and [2] to determine these quantities. The program use will be demonstrated with an example. Refer to Figure 2.1 for the microstrip geometry. When run, MICRO first clears the screen and then prints the following (italics are used to denote what the program displays, bold characters denote responses typed in by the user):

MICRO.BAS 10/27/95 08:34:23

This program calculates Zo, eff, and loss for microstrip

*enter relative dielectric constant?***2.2**

(This is the relative dielectric constant of the substrate.)

*enter dielectric thickness (cm)?***.16**

(This is the dimension, h, in Figure 2.1.)

enter conductor thickness (cm)? **.00356**

(This is t in Figure 2.1. For some substrates, the thickness is expression in ounces, for example, 1 ounce copper. One ounce corresponds to a .00356-cm thickness.)

enter frequency (Ghz)? **1.5**

enter conductivity relative to copper? **1**

(This is the relative conductivity of the metalization on the substrate.)

enter loss tangent? **.001**

(This is the loss tangent for substrate.)

enter number of line widths? **10**

(The program calculates and displays Z_0, ϵ_{re}, and α for one line at a time.

Requesting a number of widths as has just been done is useful for finding the line width when the desired impedance is known. The user has the option of stopping the calculations after any width.)

enter line width (cm)? **.4572**

(This is w in Figure 2.1.)

Now MICRO performs the calculations, clears the screen, and then displays the results.

MICRO.BAS 10/27/95 08:36:45

dielectric constant = 2.20

dielectric loss tangent = 0.0010

conductivity relative to copper = 1.0000

dielectric thickness (cm) = 0.16

line width (mils) = 0.4572

line thickness (mils) = 0.00356

ZO (ohms) = 52.364

eff = 1.860

Zo(f) (ohms) = 52.646

eff(f) = 1.866

frequency (Ghz) = 1.5000
loss (dB/cm) = 0.00406

The first two results, ZO and eff, are the quasi-static values from (2.1) to (2.8). The next two, Zo(f) and eff(f), include the effects of dispersion. At this frequency, dispersion is negligible.

enter line width (cm)   [Type 0 to exit]? **0**

(At this point, type in another line width, if desired. If another width had been input, MICRO would perform the calculations, clear the screen, and then display the results. For any set of results, the data can be printed out simply by striking the ***crtl/print screen*** keys simultaneously. By entering 0 as was done, the program jumps out of the calculation loop with the following request.)

Do another case (y or n)? **y**

If answered no (**n**), the program terminates. Either upper or lower case letters may be input. If a letter other than y or n is typed, the input is repeated. When answered yes, two options are available.

Change inputs (y or n)?

If answered yes, the entire set of inputs are requested. This is useful if the substrate or conductor parameters need to be changed. If answered no, these parameters are left unchanged and only the last two inputs (number of line widths and then the widths themselves) are requested.

Program MICENDEF finds the end-effect extension for an open-circuited microstrip line. See Figure 2.2(a). This is the length of line that must be added to the physical length to obtain the correct electrical length for an open-circuited line. When using open-circuited lines as matching stubs or to simulate an RF short, for example, the actual physical length must therefore be shortened by the end-effect extension to get the right electrical performance. Equation (2.23) is used to find the extension. As an example

MICENDEF.V10   10/27/95   09:40:40
[Current Program Date - 5/4/95]

Calculates effective length extension for a microstrip open circuit

INPUT SUBSTRATE HEIGHT (cm)? **.16**

INPUT SUBSTRATE RELATIVE DIELECTRIC CONSTANT? **2.2**

INPUT FREQUENCY? **1.5**

INPUT SMALLEST AND LARGEST LINE WIDTHS (cm)? **.381,.508**

(MICENDEF does the calculations for a range of line widths, then displays the results in tabular form.)

INPUT NUMBER OF LINE WIDTHS TO CALCULATE? **5**

MICENDEF performs the calculations, clears the screen, and then displays the results.

MICENDEF.V10 10-27-1995 09:36:22

MICROSTRIP OPEN CIRCUIT END EFFECT EXTENSION

SUBSTRATE RELATIVE DIELECTRIC CONSTANT = 2.20
SUBSTRATE HEIGHT = 0.16 (cm)
FREQUENCY = 1.500 GHz

LINE WIDTH (cm)	*LINE EXTENSION (cm)*
0.3810	*0.07767*
0.4064	*0.07896*
0.4318	*0.08018*
0.4572	*0.08135*
0.4826	*0.08246*
0.5080	*0.08352*

Note, for example, the 0.381-cm-wide line has an extension of about 0.078 cm. The next input is

PRINT OUT RESULTS (y or n)? **n**

(When answered yes, the screen is printed out.)

STORE RESULTS IN A FILE (y or n)? **y**

Answering yes to this question results in the next input.

INPUT NAME OF FILE TO STORE RESULTS? **endeff.dat**

The data is stored exactly as it was displayed on the screen. The final input is

DO ANOTHER CASE (y or n)? **n**

which allows the user to either terminate or continue.

The equivalent circuit elements for a step change in line width as shown in Figure 2.2(b) are found using program MICSTEP. The expressions in [3] are used to evaluate the two inductances and one capacitance. Step changes in width occur when two lines of different impedance are connected such as in a feed network for an array.

MSTEP.V10 10/27/95 10:23:45

Calculates Equivalent Circuit Parameters
for a Microstrip Step Change in Width

INPUT SUBSTRATE HEIGHT (cm)? **.16**

INPUT SUBSTRATE RELATIVE er AND LOSS TANGENT? **2.2, .001**

(The model was established for $\epsilon_r \leq 10$, so a flag is raised if this limit is exceeded. The user may either continue or re-enter a new dielectric constant.)

INPUT WIDER LINE WIDTH (cm)? **.4572**

INPUT NARROWER LINE WIDTH (cm)? **.2286**

(If this width is wider than the previous one, the program will not continue until a narrower width is entered.)

The program checks to see if the ratio of wide-to-narrow widths is greater than 1.5 and less than 3.5, which are the model limits. Again the user may continue or provide different widths. The next input is

INPUT DESIGN FREQUENCY (Ghz)? **1.5**

The equivalent circuit parameters are calculated, the screen cleared, and the results displayed.

MSTEP.V10   10-27-1995   10:27:02

SUBSTRATE HEIGHT = 0.16 (cm)
SUBSTRATE er = 2.20 and LOSS TANGENT = 0.0010
FREQUENCY = 1.500
WIDE LINE WIDTH = .4572 (cm)
NARROW LINE WIDTH = .2286 (cm)

L1 = 0.017 nH
Cs = 0.013 pF
L2 = 0.012 nH

L_1 is the inductor on the narrow side of the width change. The reference plane for this circuit is right at the junction. There are no additional line lengths or extensions to consider when accounting for the dimension change. As with MICENDEF, options for printing the results, storing in a file, and continuing with the program are given.

A microstrip bend is analyzed with program MICBEND, which also uses expressions from [3] to find the equivalent circuit elements. The bend and its circuit are in Figure 2.2(c). With this equivalent circuit, the reference planes are moved out toward the edge of the intersection region as shown by the dashed lines in the figure. If the electrical length of a line that includes a bend is important, the phase shift through the bend is accounted for by the equivalent circuit. The physical distance through the bend should not be counted as contributing to the electrical length if the equivalent circuit is used in the analysis. Bends are often required to route lines.

MICBEND.V10   10/27/95   11:30:12
[Current Program Date - 5/4/95]

Calculates equivalent circuit elements for an unchamfered microstrip bend

INPUT SUBSTRATE HEIGHT (cm) AND RELATIVE DIELECTRIC CONSTANT?
.16,4

(The model is valid for $2.5 \le \epsilon_r \le 15$, so the user is flagged if the ϵ_r falls outside this range. The option to continue or change the value is provided.)

INPUT SMALLEST AND LARGEST LINE WIDTHS (cm) FOR CALCULATION?
.254,.381

(Like MICENDEF, the program does a range of widths.)

If the line width-to-substrate height ratio is less than 0.1 or greater than 5, another warning is issued since these are outside the range of the model.

INPUT NUMBER OF LINE WIDTHS TO CALCULATE? **5**

The output is

MICBEND.V10 10-27-1995 11:28:47
UNCHAMFERED MICROSTRIP BEND EQUIVALENT CIRCUIT ELEMENTS

SUBSTRATE RELATIVE DIELECTRIC CONSTANT = 4.00
SUBSTRATE HEIGHT = .16 (cm)

WIDTH (cm)	SERIES IND (nh)	SHUNT CAP (pF)	SERIES IND (nH)
0.2540	0.066	0.229	0.066
0.2794	0.086	0.269	0.086
0.3048	0.105	0.313	0.105
0.3302	0.123	0.359	0.123
0.3556	0.140	0.409	0.140
0.3810	0.157	0.462	0.157

For the bend, the two inductors on either side of the capacitor are equal. Again results may be printed out, stored, and the analysis terminated or continued.

Because of the frequent occurrence of bends, much work has been done on modifications that minimize the discontinuity introduced by them. Chamfering or cutting off a portion of the corner has proven successful toward that end. Figure 2.4 shows a chamfered corner. The bend is usually called a mitered bend. Based on theoretical and experimental results, an optimum chamfer has been developed and is given by (2.20). Interestingly enough, although the optimum chamfer greatly reduces the mismatch associated with a bend, it still affects the electrical length. The electrical length through most bends is less than the physical length by the amount of (2.21). This value must be subtracted from the physical length. Again this is important when a mitered bend is used on a line that must be a certain electrical length. MICMITER calculates the optimum miter (see Figure 2.4 for a definition of dimensional terms) and the change in electrical length for that miter.

MMITER.v10 10/27/95 12:15:02
[Current Program Date - 11/1/95]

Calculates optimum chamfer for a mitered microstrip bend

INPUT SUBSTRATE RELATIVE DIELECTRIC CONSTANT? **2.2**

(Model valid for $\epsilon_r \leq 25$. Program flags this.)

INPUT SUBSTRATE HEIGHT (cm)? **.16**

INPUT LINE WIDTH (cm)? **.4572**

(For best results, the line width should be equal to or greater than four times the substrate height. Program issues a warning if line width is too narrow.)

INPUT FREQUENCY (Ghz)? **1.5**

The output is

MMITER.v10 10/27/95 12:16:32

SUBSTRATE RELATIVE er = 2.20

SUBSTRATE HEIGHT = 0.16 (cm)

LINE WIDTH = 0.4572 (cm)

FREQUENCY = 1.500000 (Ghz)

CHAMFER = 0.4880 (cm)

[Chamfer measured from outside corner of bend along outside edge of line]

REDUCTION IN ELECTRICAL LENGTH = 0.2655 (cm)
[Compared to physical length of 0.4572 (cm)]

DO ANOTHER CASE (y or n)? **n**

Note that the miter is measured along the outside edge of the joining lines. It is the length of each of the vertical and horizontal dashed lines shown at the bend in Figure 2.4. Also the figure for the reduction in electrical length is just that. To find the electrical length, it must be subtracted from the physical length.

The parameters for the microstrip T-junction are the transformers on each side of the main line, the junction susceptance, and the three reference plane extensions (see Figure 2.3). The reference plane extensions are measured from the center of the junction as seen in Figure A.1. Also shown are the widths and characteristic impedances of each branch of the junction. T-junctions are commonly used in power splitters and other portions of feed networks. MICTEE calculates all of the elements of the equivalent circuit. The

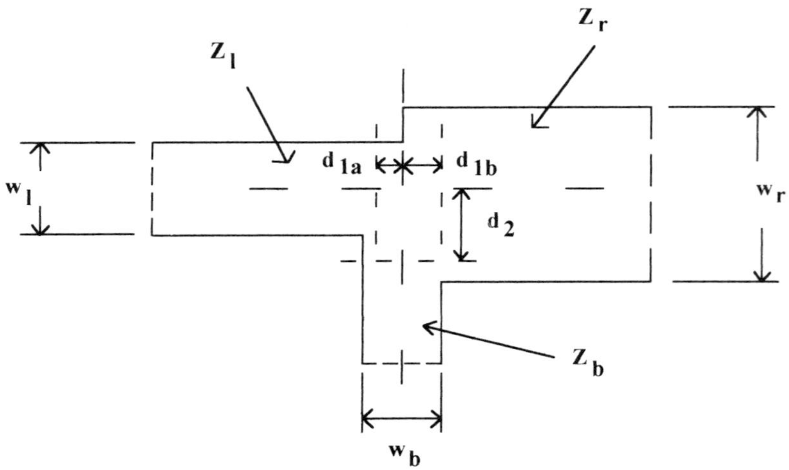

Figure A.1 Details of a microstrip T-junction.

expressions given in [3] are used for each element. When using the T-junction equivalent circuit for design, attention must be paid to the reference plane extensions as they affect the electrical length of each line connected to the junction.

MICTEE.V11 10/27/95 20:25:45
[Current Program Date - 20/4/95]

Calculates equivalent circuit elements for a microstrip T-junction

INPUT LINE IMPEDANCES (i) OR LINE WIDTHS (w)? **w**

(The calculation can be done either using the impedances of each line or their widths. This first case will use the widths.)

INPUT SUBSTRATE HEIGHT (cm)? **.16**

INPUT SUBSTRATE RELATIVE DIELECTRIC CONSTANT? **2.2**

INPUT MAIN LINE WIDTH (cm) ON LEFT SIDE? **.254**

INPUT MAIN LINE WIDTH (cm) ON RIGHT SIDE? **.4572**

INPUT BRANCH LINE WIDTH (cm)? **.127**

INPUT CONDUCTOR THICKNESS (cm)? **.00356**

INPUT START AND STOP FREQUENCIES (GHz) FOR ELEMENT CALCULATION?
1.4,1.6

(The various elements have some form of frequency dependency.)

INPUT NUMBER OF FREQUENCIES TO CALCULATE? **5**

The output is

MICTEE.V11 10-27-1995 19:55:37

Microstrip T-Junction Equivalent Circuit Elements

LEFT SIDE MAIN LINE WIDTH = 0.2540 (cm)
RIGHT SIDE MAIN LINE WIDTH = 0.4572 (cm)
BRANCH LINE WIDTH = 0.1270 (cm)
SUBSTRATE HEIGHT = 0.160 (cm) SUBSTRATE er = 2.20
CONDUCTOR THICKNESS = 0.00356 (cm)

FREQ (GHz)	d1a (cm)	d1b (cm)	nia	nib	d2 (cm)	Bt (mhos)
1.400	0.0173	0.0122	1.00	1.00	0.1822	−0.000443
1.440	0.0173	0.0122	1.00	1.00	0.1822	−0.000455
1.480	0.0173	0.0122	1.00	1.00	0.1821	−0.000467
1.520	0.0173	0.0121	1.00	1.00	0.1821	−0.000479
1.560	0.0173	0.0121	1.00	1.00	0.1820	−0.000491
1.600	0.0173	0.0121	1.00	1.00	0.1819	−0.000503

Note the relatively large reference plane extension for the branch line. If the impedances of each line are known, the option, i, at the first input may be used. The program still asks for the substrate height and relative dielectric constant. Instead of requesting line widths, the inputs are

INPUT MAIN LINE IMPEDANCE (ohms) AND EFFECTIVE er ON LEFT SIDE? **60,2**

INPUT MAIN LINE IMPEDANCE (ohms) AND EFFECTIVE er ON RIGHT SIDE? **50,2.1**

INPUT BRANCH LINE IMPEDANCE (ohms) AND EFFECTIVE er? **100,1.9**

The remaining inputs are as before. The output is

MICTEE.V11 10-27-1995 19:58:04

Microstrip T-Junction Equivalent Circuit Elements

LEFT SIDE MAIN LINE IMPEDANCE = 60.00 (Ohms)
LEFT SIDE MAIN LINE EFFECTIVE er = 2.00
RIGHT SIDE MAIN LINE IMPEDANCE = 50.00 (Ohms)
RIGHT SIDE MAIN LINE EFFECTIVE er = 2.10
BRANCH LINE IMPEDANCE = 100.00 (Ohms)
BRANCH LINE EFFECTIVE er = 1.90

FREQ (GHz)	d1a (cm)	d1b (cm)	nia	nib	d2 (cm)	Bt (mhos)
1.400	0.0143	0.0119	1.00	1.00	0.1818	−0.000496
1.440	0.0143	0.0119	1.00	1.00	0.1818	−0.000495
1.480	0.0143	0.0119	1.00	1.00	0.1818	−0.000495
1.520	0.0143	0.0119	1.00	1.00	0.1817	−0.000494
1.560	0.0142	0.0118	1.00	1.00	0.1817	−0.000494
1.600	0.0142	0.0118	1.00	1.00	0.1817	−0.000493

Although the microstrip gap and notch are not used in this book, they have several uses in microstrip circuits. The gap and its equivalent circuit are shown in Figure A.2. The gap is just as the name suggests. The line is completely severed, leaving an opening of width *s*. The reference planes for the circuit are at each end of the line, so no change in the electrical length of the line up to the gap occurs. Gaps are used for dc blocks and as capacitors in filter circuits among other things. The notch is simply a gap that does not cut completely through the line as in Figure A.3. The equivalent circuit is inductor. The reference plane is at the center of the notch as indicated by the dashed line.

MICGAP determines the circuit elements for the gap using the results in [3].

MICGAP.V10 10/27/95 9:30:12
[Current Program Date - 5/4/95]

Calculates equivalent circuit elements for a microstrip gap

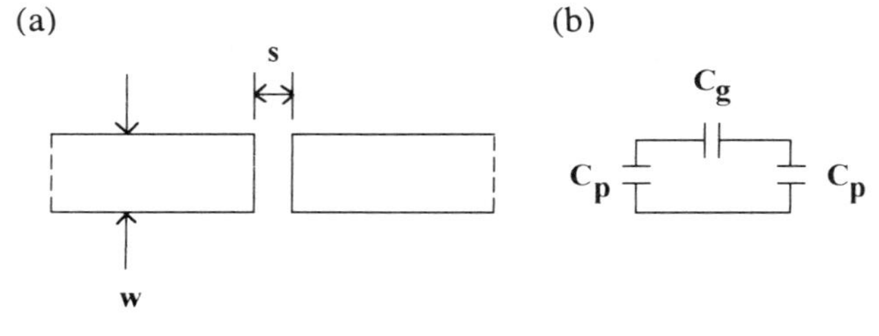

Figure A.2 Geometry and equivalent circuit for a microstrip gap: (a) microstrip gap discontinuity and (b) equivalent circuit.

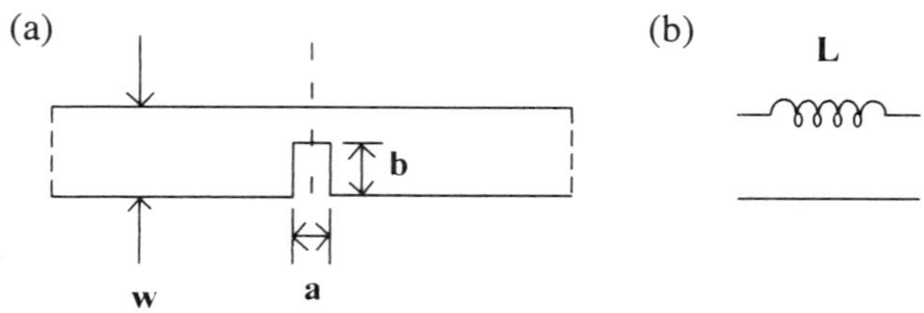

Figure A.3 Geometry and equivalent circuit for a microstrip notch: (a) microstrip notch discontinuity and (b) equivalent circuit.

INPUT SUBSTRATE HEIGHT (cm) AND RELATIVE DIELECTRIC CONSTANT?
.16,4

(The model validity limits are $2.5 \leq \epsilon_r \leq 15$. The program flags any input outside this range.)

INPUT MICROSTRIP LINE WIDTH (cm)? **.3048**

(Model limits for the width-to-substrate height ratio are $0.5 \leq w/h \leq 2$.)

INPUT SMALLEST AND LARGEST GAPS (cm) FOR CALCULATION?
.0508,.1778

(Model limits for gap-to-width ratio are $0.1 \leq s/w \leq 1$.)

INPUT NUMBER OF GAP WIDTHS TO CALCULATE? **5**

The output is

MICGAP.V10 10-27-1995 21:30:23

MICROSTRIP GAP EQUIVALENT CIRCUIT ELEMENTS
SUBSTRATE RELATIVE DIELECTRIC CONSTANT = 4.00
SUBSTRATE HEIGHT = 0.160 (cm)
MICROSTRIP LINE WIDTH = 0.3048(cm)

GAP (cm)	SHUNT CAP (pF)	SERIES CAP (pF)	SHUNT CAP (pF)
0.0508	0.001	0.073	0.001
0.0762	0.002	0.062	0.002
0.1016	0.003	0.054	0.003
0.1270	0.003	0.049	0.003
0.1524	0.004	0.045	
0.004	0.1778		
0.004	0.042	0.004	

Options are available to print out and store the results in a file.

Program MICNOTCH is used for the notch inductance using an equation presented in [3].

MICNOTCH.V10 10/27/95 10:09:46
[Current Program Date - 5/4/95]

Calculates equivalent circuit elements for a microstrip notch

INPUT SUBSTRATE HEIGHT (cm) AND RELATIVE DIELECTRIC CONSTANT?
.16,2

INPUT FREQUENCY (Ghz)? 1.5 {usl} INPUT LINE WIDTH (cm)? **.4572**

INPUT SMALLEST AND LARGEST NOTCH DEPTHS (cm) FOR CALCULATION? **.0508,.1778**

(This is the dimension, *b*, in Figure A.3. The model limits for the depth-to-line width ratio is $0 \leq b/w \leq 0.9$.)

INPUT NOTCH WIDTH (cm)? **.0127**

(This is the dimension, *a*, in Figure A.3. Model limits for notch width-to-substrate height is $a \leq h$.)

The output is

MICNOTCH.V10 10-27-1995 21:32:43

MICROSTRIP NOTCH EQUIVALENT CIRCUIT ELEMENT

SUBSTRATE RELATIVE DIELECTRIC CONSTANT = 2.20
SUBSTRATE HEIGHT = 0.160 (cm)
LINE WIDTH = 0.4572 (cm)
NOTCH WIDTH = 0.0508 (cm)
FREQUENCY = 1.500 GHz

NOTCH DEPTH (cm)	SERIES IND (nh)
0.0508	0.016
0.0762	0.035
0.1016	0.063
0.1270	0.099
0.1524	0.144
0.1778	0.197

Again this can be printed out and stored in a file.

References

[1] Gupta, K. C., R. Garg, and I. J. Bahl, *Microstrip Lines and Slotlines*, Norwood, MA: Artech House, 1979.

[2] Kirschning, M., and R. H. Jansen, "Accurate Model for Effective Dielectric Constant of Microstrip With Validity up to Millimeter-Wave Frequencies," *Electron Lett.*, Vol. 18, No. 6, 1982, pp. 272–273.

[3] Chang, K., ed., *Handbook of Microwave and Optical Components*, Vol. 1, New York, NY: John Wiley & Sons, 1989, pp. 96–107.

Appendix B
Computer Programs Supplied on Disk

The computer programs discussed in this book are available on an accompanying disk. The programs are contained in file *patchprg.exe*, which is a self-extracting, compressed file. Place the file in a directory of your choice and execute. This will create the 33 individual executable files discussed in the book. The programs are compatible with all versions of Windows. Another file, *manual.exe*, contains a user's manual. Also run it to decompress the manual. The manual is a MICROSOFT WORD for WINDOWS document and is about 117 pages long.

The following is a listing of the individual programs.

Chapter 2

micbend (microstrip bend)
micendef (microstrip open end effect extension)
micgap (microstrip gap)
micmiter (optimum microstrip bend miter)
micnotch (microstrip notch)
micro (microstrip Z_0, ϵ_{re}, and attenuation)
micstep (microstrip step width change)
mictee (microstrip T-junction)

Chapter 3

chwpatch (radiation patterns for dielectric covered patch)
cirpat (radiation patterns for circular patch)
cpatch (circular patch design)
hwpatch (radiation patterns for rectangular patch)

patch9 (rectangular patch design)
patchc (dielectric covered rectangular patch design)
patchd (rectangular and circular patch analysis)
scpatch (quarterwave, short-circuited patch design)
wrappat (wraparound patch design)
wraprnd (radiation patterns for wraparound patch)

Chapter 4

cppatch (single-feed circularly polarized patch design)

Chapter 5

capmat (capacitor matching for thick patch)
tlpatch (patch with two dielectric layers underneath)
mfpatch (impedance of stacked patches)
parapat (parasitically coupled patch design)

Chapter 6

aperdist (determines aperture distribution for an array)
arraycal (analyzes beamwidth and directivity for array)
condcal (determines traveling wave array conductances)
narrayd (radiation patterns for arbitrary array)
patchcom (patch parameters for series fed arrays)
resfeed (determines resonant array conductances)

Chapter 7

covmic (for example, Z_0 for dielectric covered microstrip)
cps (Z_0 and ϵ_{re} for coplanar strips)
cpw2 (Z_0 and ϵ_{re} for coplanar waveguide)
slotline (Z_0 and ϵ_{re} for slotline)

About the Author

Robert Sainati received a B.S.E.E. and M.S.E.E. from the University of Illinois, Urbana, IL. He also received a Ph.D. in electrical engineering from the University of Connecticut, Storrs, CT. For over twelve years, he was employed at the U.S. Navy Underwater Systems Center in New London, CT. There he developed antennas for submarine applications and performed propagation research studies and microwave systems studies. He then joined the Honeywell Defense Systems Division, Hopkins, MN, where he was involved in the analysis, design, and development of millimeter-wave printed circuit antennas. After several years there, he became a staff member at the Honeywell Corporate Technology Center, Bloomington, MN. At Honeywell he was involved in the analysis and design of electrical interconnects and packaging for high-speed Gallium Arsenide digital integrated circuits. He also was program manager of an ongoing research effort to develop high-speed interconnects and packaging. Later, he joined the Gallium Arsenide millimeter-wave integrated circuit group where he investigated the application of integrated circuit technology in millimeter-wave phased arrays and radars.

In 1990 Honeywell spun off the majority of its defense business into a company named Alliant Techsystems. Dr. Sainati joined the Research and Technology Center of Alliant in Hopkins, MN. At Alliant, he was responsible for the design and development of advanced printed circuit antenna concepts including phased arrays at millimeter-wave frequencies. He led several in-house research and external contract programs. In late 1993, he joined the CSIRO Division of Radiophysics in Sydney, Australia. He participated in a research program aimed at developing antenna systems for 60 GHz broadband wireless LANs. He has since rejoined Alliant Techsystems where he is currently developing several millimeter-wave printed circuit antennas.

Dr. Sainati is a member of the IEEE including the Antennas and Propagation and Microwave Theory and Techniques Societies within that organization. He has published approximately twelve papers in refereed journals and given about six presentations at various conferences. Dr. Sainati holds one United States patent.

Index

The Artech House Antenna Library

Helmut E. Schrank, *Series Editor*

Advanced Technology in Satellite Communication Antennas: Electrical and Mechanical Design, Takashi Kitsuregawa

Analysis Methods for Electromagnetic Wave Problems, Volume 2, Eikichi Yamashita, editor

Analysis of Wire Antennas and Scatterers: Software and User's Manual, A. R. Djordjević, M. B. Bazdar, G. M. Bazdar, G. M. Vitosevic, T. K. Sarkar, and R. F. Harrington

Analysis Methods for Electromagnetic Wave Problems, E. Yamashita, editor

Antenna-Based Signal Processing Techniques for Radar Systems, Alfonso Farina

Antenna Design With Fiber Optics, A. Kumar

Broadband Patch Antennas, Jean-François Zürcher and Fred E. Gardiol

CAD for Linear and Planar Antenna Arrays of Various Radiating Elements: Software and User's Manual, Miodrag Mikavica and Aleksandar Nešić

CAD of Aperture-fed Microstrip Transmission Lines and Antennas: Software and User's Manual, Naftali Herscovici

CAD of Microstrip Antennas for Wireless Applications, Robert A. Sainati

The CG-FFT Method: Application of Signal Processing Techniques to Electromagnetics, Manuel F. Cátedra, Rafael P. Torres, José Basterrechea, Emilio Gago

Electromagnetic Waves in Chiral and Bi-Isotropic Media, I.V. Lindell, S.A. Tretyakov, A.H. Sihvola, A. J. Viitanen

Fixed and Mobile Terminal Antennas, A. Kumar

Generalized Multipole Technique for Computational Electromagnetics, Cristian Hafner

Handbook of Antennas for EMC, Thereza Macnamara

Integral Equation Methods for Electromagnetics, N. Morita, N. Kumagai, and J. Mautz

IONOPROP: Ionospheric Propagation Assessment Program, Version 1.1: Software and User's Manual, Hernert V. Hitney

Four-Armed Spiral Antennas, Robert G. Corzine and Joseph A. Mosko

Introduction to Electromagnetic Wave Propagation, Paul Rohan

Introduction to the Uniform Geometrical Theory of Diffraction, D. A. McNamara

Microwave Cavity Antennas, A. Kumar and H. D. Hristov

Millimeter-Wave Microstrip and Printed Circuit Antennas, Prakash Bhartia

Mobile Antenna Systems, K. Fujimoto and J. R. James

Modern Methods of Reflector Antenna Analysis and Design, Craig Scott

Moment Methods in Antennas and Scattering, Robert C. Hansen, editor

Monopole Elements on Circular Ground Planes, M. M. Weiner et al.

Near-Field Antenna Measurements, D. Slater

Passive Optical Components for Optical Fiber Transmission, Norio Kashima

Phased Array Antenna Handbook, Robert J. Mailloux

Polariztion in Electromagnetic Systems, Warren Stutzman

Practical Phased-Array Antenna Systems, Eli Brookner et al.

*Practical Simulation of Radar Antennas and Radome*s, Herbert L. Hirsch and Douglas C. Grove

Radiowave Propagation and Antennas for Personal Communications, Kazimierz Siwiak

Reflector and Lens Antennas: Software User's Manual and Example Book, Carlyle J. Sletten, editor

Reflector and Lens Antenna Analysis and Design Using Personal Computers, Carlyle J. Sletten, editor

Shipboard Antennas, Second Edition, Preston Law

Small-Aperture Radio Direction-Finding, Herndon Jenkins

Spectral Domain Method in Electromagnetics, Craig Scott

Understanding Electromagnetic Scattering Using the Moment Method: A Practical Approach, Randy Bancroft

Waveguide Components for Antenna Feed Systems: Theory and CAD, J. Uher, J. Bornemann, and Uwe Rosenberg